St Antony's Series
Series Standing Order ISBN 978-0-333-71109-5
(*outside North America only*)

You can receive future titles in this series as they are published by placing a
standing order. Please contact your bookseller or, in case of difficulty, write to us
at the address below with your name and address, the title of the series and the
ISBN quoted above.

Customer Services Department, Macmillan Distribution Ltd, Houndmills, Basingstoke,
Hampshire RG21 6XS, England

The Rise and Fall of Privatization in the Russian Oil Industry

Li-Chen Sim
Assistant Professor, Zayed University, United Arab Emirates

In Association with
Palgrave Macmillan

First published 2008 by
PALGRAVE MACMILLAN

Palgrave Macmillan in the UK is an imprint of Macmillan Publishers Limited,
registered in England, company number 785998, of Houndmills, Basingstoke,
Hampshire RG21 6XS.

Palgrave Macmillan in the US is a division of St Martin's Press LLC,
175 Fifth Avenue, New York, NY 10010.

Palgrave Macmillan is the global academic imprint of the above companies
and has companies and representatives throughout the world.

Palgrave® and Macmillan® are registered trademarks in the United States,
the United Kingdom, Europe and other countries.

ISBN-13: 978-0-230-20298-6   hardback

This book is printed on paper suitable for recycling and made from fully
managed and sustained forest sources. Logging, pulping and manufacturing
processes are expected to conform to the environmental regulations of the
country of origin.

A catalogue record for this book is available from the British Library.

Library of Congress Cataloging-in-Publication Data

Sim, Li-Chen, 1972-
   The rise and fall of privatization in the Russian oil industry /
   Li-Chen Sim.
       p. cm. — (St Antony's series)
   Includes bibliographical references and index.
   ISBN-13: 978-0-230-20298-6 (alk. paper)
   1. Petroleum industry and trade—Russia (Federation)
   2. Privatization—Russia (Federation)   3. Industrial policy—Russia
   (Federation)   I. Title.

   HD9575.R82S458 2008
   338.2'7280947—dc22   2008016422

10   9   8   7   6   5   4   3   2   1
17   16   15   14   13   12   11   10   09   08

Transferred to Digital Printing 2011

For my parents,
Mike and Anne

Contents

Acknowledgements

When I first embarked on my research, I had little idea of what I was letting myself in for – the mountains of material to sieve through, the periods of self-doubt, the stages of writer's block. The fact that this book has seen the light of day is a testament to people who have helped me along the way, and to whom I owe a great debt of gratitude.

My thanks go out to my mentor, Prof. Archie Brown, for his unflagging optimism that the project would one day be published. By playing the devil's advocate – which sometimes kept me up at night thinking of a suitable rebuttal – he forced me to examine, refine and strengthen my arguments and to see the merits of alternative perspectives. His generosity with his time and advice as well as his personal kindness are much appreciated. Dr Paul Chaisty and Prof. Thane Gustafson, who read an earlier draft of the manuscript, kindly shared with me invaluable suggestions on getting my work ready for publication. However, none of them bear any responsibility for errors of fact or judgement.

Research for this book was conducted over a number of years in both the United Kingdom and Russia. I am greatly indebted to persons I have interviewed in the course of my research. They were without exception generous of their time, forthcoming with information and did not shy away from dealing with sensitive and awkward issues concerning oil privatization, even when some interviews were conducted in the midst of the YUKOS affair. It was certainly a real privilege for me to engage with participants who have directly influenced the policy process in Russia. I would also like to acknowledge the generosity of organizations such as the Oxford Institute of Energy, the Bodleian Library (University of Oxford), the Russian & East European Studies Centre at St Antony's College (University of Oxford) and the *Kommersant'* group (Moscow), which made available their holdings on the topic; and consultancy firms such as Troika Dialog, Renaissance Capital and Aton, which did not hesitate to share their well-researched reports on the Russian oil industry.

· Most important of all, I wish to thank my parents, Mike and Anne, to whom this book is dedicated, for their unstinting and constant support in everything that I do; and my husband, Frank, for his love, encouragement, humour (when it was most needed) and patience.

Notes on Transliteration and Translation

This book uses the transliteration system adopted in *The Cambridge Encyclopedia of Russia and the Former Soviet Union* and *Post-Soviet Affairs* whereby:

e   = e (except at the beginning of a word or name, such as Yevgeniy)
й   = y
ы   = y
ь   = '
ю = yu
я   = ya

When referring in the text to a work published in English by a Russian author, the name will be transliterated according to the above system (for example, Yuliya Latynina, Yegor Gaydar). However, in the footnotes and bibliography, the name will be cited as it appears in the original language of the source (for instance, Julia Latynina, Yegor Gaidar). An exception is made if the subject is well known in the English-speaking world and uses a modified transliteration of the name. Hence, the book refers to LUKoil and not LUKoyl, and to Rair Simonyan not Rayr Simonyan.

All translations from Russian sources are by the author, except where otherwise indicated.

Abbreviations

GKI — Gosudarstvennyy Komitet Imushchestva
State Committee for the Management of State Property; later upgraded to a full ministry (Ministry of State Property) and now renamed the Ministry of Property Relations

IMF — International Monetary Fund

Mintopenergo — Ministerstvo Topliva i Energetiki
Ministry of Fuel and Energy; it no longer exists as a separate ministry and is merged into the Ministry of Industry and Energy

VIOC — Vertically integrated oil company

Principal Political and Economic Actors in the Book

Name	Designation	Organization
Alekperov, Vagit	President	LUKoil
Berezovskiy, Boris	Co-owner (former)	Sibneft'
Bogdanchikov, Sergey	President	Rosneft'
Chernomyrdin, Viktor	Prime Minister (former)	Russian Federation
	Chairman (former)	Gazprom
Chubays, Anatoliy	Chief Executive	Unified Energy Systems
	Chairman (former)	GKI
Fomin, Anatoliy	President (former)	Slavneft'
Gaydar, Yegor	Acting Prime Minister (former)	Russian Federation
Gutseriev, Mikhail	President (former)	Slavneft'
	Deputy Speaker (former)	Duma
Khodorkovskiy, Mikhail	President (former)	Menatep
	Chief Executive (former)	YUKOS
Lopukhin, Vladimir	Minister (former)	Mintopenergo
Muravlenko, Sergey	President (former)	YUKOS
Putilov, Aleksandr	President (former)	Rosneft'
Putin, Vladimir	President (former)	Russian Federation
Shafranik, Yuriy	Minister (former)	Mintopenergo
Yel'tsin, Boris	President (former)	Russian Federation

1
Introduction to the Policymaking Process

The transformation of the oil industry is one of the most significant changes that have occurred as part of the transition to a market-based economy in post-Soviet Russia. At the beginning of 1992, there were 300 fully state-owned oil entities ranging from the LUKoil business *kontsern*, production associations, enterprises involved in exploration, refining and distribution as well as research centres.[1] Within a decade, six privately owned oil companies emerged to dominate the Russian oil industry, namely, LUKoil, YUKOS, Surgutneftegaz, Sibneft', Tyumen' Oil (TNK) and Slavneft'. They have been ranked among the 50 largest oil companies in the world, and at the end of 2003, they collectively accounted for just over 75% of Russia's total oil output, 70–80% of oil exports and 60% of the country's proven oil reserves. Comparatively, the remaining three state-owned oil majors – Rosneft', Tatneft' and Bashneft' – accounted for only 8% of Russia's annual oil production and 15% of reserves in 2003.[2] With the sale of Yuganskneftegaz, a key oil production subsidiary of YUKOS, and that of Sibneft' in December 2004 and September 2005, respectively, the share of oil produced by state-owned companies has since risen to one-third of Russia's total production today. Nevertheless – and this is an often overlooked point – two-thirds of Russian oil production remains in the hands of privately owned oil companies.[3]

The privatization of the Russian oil industry, particularly during the heyday of 1992–2003, is intriguing because it is contrary to the inherited and contemporary trends within the oil industry. First, the pivotal role of oil is not merely a post-Soviet phenomenon, but rather underpinned much of the development in tsarist and Soviet economies; it is therefore surprising that such a historically strategic and valuable commodity was allowed to be privatized. Tsarist Russia, for instance, was one of the largest producers of oil in the world at the turn of the twentieth century, and oil was the second-largest contributor to the country's export earnings after grain.[4]

In the case of the Soviet Union, it was basically a 'one crop economy' as far as hard currency receipts were concerned.[5] Between 1971 and 1975, oil exports accounted for 28% of Soviet hard currency earnings per annum on average, a share which increased to 46% during 1976–1980 (or US$7.1 billion) and then to 60% (or US$12.7 billion) during 1981–1985. In comparison, gas exports comprised only 1.4%, 7.6% and 17% of hard currency earnings during the respective five-year periods.[6] Admittedly, the value of exports of arms and military equipment amounted to some US$20 billion per year on average; however, such sales were mainly to Soviet satellite states that either did not pay in cash or did not pay at all. In 1989, for instance, the Soviet Union exported approximately US$21.8 billion worth of arms but received only US$1.7 billion in cash. Actual foreign currency receipts from the arms trade in the 1980s have been estimated at US$2–2.5 billion a year on average.[7] Hence, to remark that the state-owned oil industry essentially 'guaranteed the resolution of social and economic problems of the country for several decades',[8] by subsidizing domestic consumption and the growth of manufacturing industries, was not an exaggeration. This was because 'had there been no Samotlor oil, events would have forced [the Soviet Union] to start economic restructuring 10 or 15 years earlier'.[9] In fact, the decline in Soviet oil production and falling world prices for oil in the second half of the 1980s have been commonly cited as key reasons for the Soviet Union's poor economic performance and dissolution.[10]

In today's Russia, oil continues to be an invaluable source of revenue. It comprises at least 25% of federal budget revenues and 40% of export earnings.[11] With oil accounting for 18% of gross domestic product (GDP), the oil price boom and its effects on production and export levels in Russia have also fuelled much of the country's economic growth since 2000, where every 10% increase in the price of oil results in an increase of 0.7% in GDP growth.[12] In comparison, other export sectors such as gas and weapons sales are much less significant to the economy, accounting for shares of 15% and 3.8%, respectively, of total export revenues between 2000 and 2004.[13] In this connection, it is arguable that 'Russia's stability is premised on the high price of oil and the high personal popularity of [then President] Vladimir Putin'.[14]

Second, the rapid pace with which Russia's oil industry was demonopolized, corporatized and privatized was in stark contrast to developments in other energy subsectors. In the coal industry, for example, the former Soviet Ministry of Coal was abolished in favour of a new corporatized entity known as Rosugol'. However, Rosugol' retained many of the functions and the structure of its predecessor: it allocated and distributed subsidies to mines and made decisions regarding equipment replacement, mine reconstructions and shipment orders. Rosugol' was liquidated only at the end of 1997, which ushered in a period of privatization within the coal industry.[15] Gazprom, the former Soviet Ministry of Gas, was corporatized in mid-1992, but the state's shareholding of 38% remained unchanged for 15 years until December 2005.[16] The company continues to control the entire gas production and transportation system in Russia and has repeatedly warned that any attempt to break up Gazprom's monopoly would result in dire social, safety and economic consequences.[17]

The privatization of the Russian oil industry is also intriguing for a third reason: there has been no compelling rationale to do away with state-ownership in the first place since there appeared to be little empirical evidence of any direct correlation between ownership type and a world-class oil company. For instance, the 20 largest oil companies in the world in terms of reserves, output and revenues include an almost even mixture of state and privately owned companies in each category.[18] Moreover, state-ownership is the norm in most oil-rich countries, particularly in the developing world, where a single, state-owned company owns and manages the oil industry in Saudi Arabia (Aramco), Venezuela (Petróleos de Venezuela) and the United Arab Emirates (Abu Dhabi National Oil Company).[19] Within the industrialized world, only the United States of America, the United Kingdom, Canada and France have a privately owned oil industry, whereas the state is still a significant shareholder in the Norwegian and Italian oil sectors. For Russia, therefore, international experience offered no clear consensus on the merits of privatizing its oil industry.

Attempts to explain the privatization of the Russian oil industry despite the above considerations typically centre on the self-serving pecuniary interests of tycoons from the private sector. Corporations run by post-Soviet tycoons have allegedly 'in large part determined the foreign and domestic policy of Russia [and] influenced the choice of economic priorities'.[20] Putin has likewise accused companies in the natural resource sector of being prepared to sell their oil and gas resources, thereby 'boosting their capitalization at the nation's expense'.[21] As Philip Hanson and Elizabeth Teague have observed, 'the dominant paradigm in analyses of Russian development [is] one of state assets ruthlessly plundered by a tiny, grasping elite of business "oligarchs" who had "captured" the state and taken over its functions'.[22]

In this connection, I will examine the role of business in the privatization of the Russian oil industry from 1992 to 2006 and consider the extent to which its leaders were the driving force behind the process. The contribution of this book to scholarly literature lies in the intention to diverge somewhat from the tendency to concentrate on post-Soviet tycoons and their companies and to explore as well the influence on policymaking exercised by relatively less well-known representatives of the business community, namely, the Soviet-era general managers of oil companies.[23] Indeed, while acknowledging that the latter were 'the most powerful of Russia's new capitalists' and the 'canniest players in the fight for [Russia's] loot', the study by Chrystia Freeland is, nevertheless, limited to the roles played by two of the most well-known representatives of the genre, Vagit Alekperov and Viktor Chernomyrdin. In addition, there is a general assumption that a wide gulf exists in the leadership traits of top managers in the more industrialized countries and leading representatives of the so-called red directors corps. For example, empirical studies conducted in the early 1990s identified Russian management culture to be characterized by a high tolerance for inequality in business relationships, a high tendency to avoid risk and a high level of appreciation for social collectivism, all of which appeared to be diametrically opposite to management values in the West.[24] As a result, it was feared that for Russia, 'the prospects of denationalization, redistribution of decision-making authority for enterprise performance, and ultimately the privatization of state-controlled enterprises [were] overwhelming'.[25] Through the case studies of particular oil companies, this book will attempt to evaluate the extent to which Soviet-era general managers are as rational as their Western counterparts. Is their behaviour conditioned by cultural imperatives or by the prevailing structure of incentives and institutions? How successful have they been in adapting to the new post-Soviet economic realities? Given that the decision to privatize the oil industry was in fact adopted at the end of 1992, and hence prior to the emergence of big business, what role did these 'red directors' play in the policy process?

In addition, analysing the role of Soviet-era general managers may shed some light on the relationship between legacies and future success, particularly the extent to which membership in the Soviet *nomenklatura* conferred advantages when operating in the new post-Soviet environment. To what extend did being part of this *nomenklatura* shape their decisions vis-à-vis oil exploration, support for social programmes for oil workers, transparency, payment of taxes and relations with the state? Also, how far did new opportunities and new incentives affect their decisions and behaviour pertaining to the privatization of the Russian oil industry?

Moreover, given the crucial role that oil plays in Russia, there is a surprising lack of scholarly literature focusing on the role of oil companies and their representatives in specific aspects of domestic policy.[26] Exceptions include studies on the evolving structure of the oil industry, ties between banks and oil companies, profiteering within the industry, the composition and perceptions of the oil elite and LUKoil's influence in the political arena.[27] The focus of this book – on oil companies and the impact of their leaders on policies related to oil privatization in Russia – therefore will be to address this gap in the literature.

Business as a privileged group

In a seminal work that initiated a debate on the nature of business in politics, Charles Lindblom argued that business is the most privileged of all groups that seek to influence public policy, claiming that in a market-based economy, 'no other group of citizens can compare with businessmen, even roughly, in effectiveness'.[28] In Russia, rich and politically well-connected businessmen comprised 'by far the most influential lobbies'[29] during the administration of former president Boris Yel'tsin and 'became an essential component of the political system ... it [was] hard to imagine a Russian president being able to rule without their support'.[30] Putin himself acknowledged the influence of business on the state in remarks to Russia's leading businessmen that 'you have yourselves formed this state, to a large extent through political and quasi-political institutions under your control'.[31]

There are various reasons why big business is particularly influential in the policymaking process. The first concerns the structural power of business whereby the rules of the game, which tend to favour one group over another, are built into the political system. According to Lindblom, 'because public functions in the market system rest in the hands of businessmen, it follows that jobs, prices, production, growth, the standard of living and the economic security of everyone all rest in their hands'.[32] This structural advantage means that business does not even need to be organized to influence public policy. Indeed, a Russian state official acknowledged that leading businessmen 'control hundreds of companies and employ millions of people ... At the end of the day, we have to work with these people. We don't simply want them to go bust'.[33]

Likewise, British Petroleum admitted that the sheer size of its merger with TNK was crucial because 'in Russia, if you are a tiny company, you are irrelevant to the government. But if you're huge, then you're more important and it is more difficult for unpleasant things to happen to you. They can still happen, the risk is not completely eliminated ... But any nasty thing that happens would become a very big political issue'.[34]

Second, big business derives political influence from its role as a corporation in view of its financial and organizational resources as well as access to policymaking circles. Large companies possess extraordinary sources of funds; in Russia, for instance, the combined turnover of the top 12 privatized companies is roughly equal to the federal government's annual revenues.[35] This financial power has enabled business to influence the political process by buying votes in the Duma and funding political campaigns.[36] Indeed, it is widely acknowledged that Yel'tsin's re-election campaign in 1996 was financed by a coterie of rich businessmen[37] who also made available to him other corporate resources such as 'their media bullhorns, their talented staff'.[38]

In addition, leading businessmen enjoy ease of access to government officials as a result of the movement of government personnel into corporate positions and vice versa.[39] Vagit Alekperov, president of LUKoil, was formerly the first deputy oil minister of the Soviet Union in 1990. This, coupled with the fact that other LUKoil employees have previously been appointed to key government energy posts,[40] is one reason why the company was able to influence the privatization of the oil industry, in particular, the establishment of vertically integrated oil companies (henceforth VIOCs) and limiting the amount of LUKoil shares available for use as collateral in the loans-for-shares scheme. Under Putin, the Russian Union of Industrialists and Entrepreneurs (RSPP), which includes representatives of big business among its members, enjoyed 'insider status' at least up until the YUKOS affair of 2003–2004. Indeed, the RSPP used to meet regularly with the president, and its members were part of working groups set up by the government to introduce corporate governance guidelines and other pro-business legislation and policies.[41]

A third explanation why big business is a particularly influential organized interest lies in its relatively limited membership size. Smaller groups have a greater likelihood of engaging in successful collective action than larger groups because the former can organize with less delay and agree on mutual advantages more easily.[42] In the case of the loans-for-shares deal in Russia, through which valuable state-owned companies were sold at low prices to a small group of private owners, the latter 'reached an agreement on who would take what. We agreed not to get in each others' way ... [Vladimir] Potanin would get the best company Norilsk Nickel, because he had come up with the plan in the first place. [Mikhail] Khodorkovskiy would get YUKOS'.[43]

For many observers, this pervasive influence of corporate groups on public policy is undesirable since narrowly focused groups seek only to increase their respective shares of national income, instead of increasing national wealth *per se*. As such, they 'interfere with an economy's capacity to adapt to change and to generate new innovations'.[44] For instance, in post-communist countries where reform efforts were piecemeal and inconsistent, the early winners of initial reforms succeeded in blocking further economic liberalization through the capture of state institutions; they were determined to maintain and defend the resulting partial reform equilibrium, since further reform would have eroded the distortions on which their initial gains were based.[45]

According to an alternative perspective, however, big business in Russia has been a positive force for change. As a result of the loans-for-shares auctions, leading privately owned business groups took control of hitherto insider-dominated companies that had been reluctant to embrace much-needed structural reforms. Having amassed highly concentrated shares in valuable companies often through illegal means, 'it [was] no longer in their interest to violate the law and engage in rent-seeking that destabilises the macroeconomy'.[46] Instead, business tycoons have supported the need for a rule-based market environment in order to attract foreign investment, which will in turn enhance the value of their own shares in these companies. This is because 'Russian businessmen are rational people. The change in outlook may be cynical. But it's rational: honesty is profitable now ... We all understand that. We're not idiots.'[47]

In any case, to acknowledge the power (positive or negative) of leading corporate groups is not to imply that business preferences always prevail. This is because the possession of overwhelming power resources by big business does not directly result in the disproportionate exercise of power – such resources have to be effectively converted into political power.[48] As such, 'in no country has business been unchallenged'[49] as evidenced by increasing government regulation over business and the latter's conflicts with environmental, animal welfare activists and other consumer groups.

The most important factor that shapes the extent of business influence on public policy is, undoubtedly, the state. First, the state can influence the formation and strength of business groups. In Russia's case, the influence of corporate groups on public policy was largely the result of initial and crucial decisions by leading state officials to create a constituency supportive of privatization and market reforms. Suffice to note for now that these state officials recognized that the existing institutional constraints (or 'stakeholders') at the outset of the reform process – the state, branch ministries, managers, workers and regional authorities – had to be systematically 'appeased, bribed or disenfranchised'.[50] At the same time, given that there was 'no clear pro-reform majority'[51] in the government or society, the team of so-called young reformers, led by acting prime minister Yegor Gaidar, were determined to create a property-owning class through privatization. In this connection, a Russian economic expert recalled a conversation with one of the 'young reformers', Anatoliy Chubays, in which the latter 'realise[d] that the owners will mostly be criminally oriented people. But, he said, there are no others now . . . And without real entrepreneurs and owners, Russia will never get out of the hole'.[52]

Second, the state can influence the orientation of business groups. For instance, during the mid-1990s, the state designed incentives for commercial banks to achieve their profit-maximizing interest by giving up support for destabilizing, inflationary policies, in exchange for large stakes in valuable companies, which would offer high returns only in a stable economic environment.[53] Putin also attempted to orientate big business away from an overt role in politics by taking punitive actions against the most outspoken and ambitious tycoons and by only holding discussions with leading businessmen within an organizational setting instead of on an individual basis as was the case previously.[54]

Third, the type of arena provided by the state also affects the extent to which business can exert its influence. The more fragmented the political arena, the more channels there are for business to exert influence on public policy. This is because the state is not a homogeneous, faceless whole, but is made up of different bureaucratic agencies with distinct aims and competing leaders who try to influence policymaking with regard to oil privatization and who achieve varying degrees of success in so doing. Under Yel'tsin, bureaucratic rivalries within the state apparatus were especially pronounced and exploited by business groups, as will be evident later.[55] This led some observers to remark that Yel'tsin's Russia was a 'government of lobbyists',[56] since 'some ministers or deputy prime ministers served as open lobbyists for the interest group they supposedly regulated'.[57] For example, Yuriy Shafranik, head of Mintopenergo, admitted that he had collaborated with oil and other energy companies to 'create our own "energy lobby"' in parliament[58] to 'ensure the loyalty of future deputies and factions to the fuel and power complex'.[59]

The foregoing discussion on the role of business in politics, and in particular the dynamic relations between big business and the state in Russia, is perhaps best summed up by Stephen Sestanovich's observation that 'of all the potential forces in Russian politics, "money" has the strongest material base and the greatest doubts about its own legitimacy'.[60]

Analytical framework: Rational choice institutionalism

Underlying much of the discussion in this book is an approach known as 'rational choice institutionalism'.[61] This analysis on the interaction between human agency and institutional parameters has been used to explain a variety of transition-related issues in Russia, including rent-seeking behaviour by the tycoons, the creation of the superpresidential system, institutional innovations to increase the acceptability of privatization and abuse of minority shareholders.[62]

The following are the assumptions of rational choice institutionalism for the purpose of this book. First, the individual is the basic unit of analysis. Political actors make decisions and determine policy outcomes, not institutions, which can even be destroyed or recrafted by the former.[63] This is in line with an 'agency-centred' view of politics.

Second, the political actor is self-interested. As Chapter 2 will explain, this self-interest can manifest itself in different forms – power, budgetary allocation, material rewards or ideational mission – depending on the actor in question, since actors 'need not be driven by pecuniary interests – the value they maximize is not necessarily economic'.[64] In this book, I will adopt a 'thin' notion of rationality by focusing on each actor's overriding interest to avoid tautological explanations; indeed, 'if we make our motivational assumptions complicated enough, we can "explain" any kind of behaviour – which of course means that we are explaining absolutely nothing ... We must definitely resist the temptation of postulating more than a very few basic motives in our theory, whether for the sake of "greater realism" or for any other reasons'.[65]

Third, the self-interested actor is also intendedly rational, so that he attempts to maximize the possibility that his highest-ranked preference is achieved. However, given that politics does not take place in a vacuum and is contextual in nature, his options are limited by a set of positive incentives and negative inducements embedded in institutions. His rationality is therefore 'bounded' by institutions, so that more often than not, he does not actually achieve his optimal preference, but adjusts to an outcome that is 'good enough'. This is known as 'satisficing' behaviour.[66]

Finally, institutions comprise formal rules and legislation, as well as informal constraints such as conventions and behavioural norms, all of which 'define repertoires of more or less acceptable courses of action that will leave considerable scope for the strategic and tactical choice of purposeful actors'.[67] Institutions therefore structure interactions between actors by delineating the permissible range of alternative solutions to a given issue and thus provide useful and predictable outcomes of how other participants in the policy process would act. Nevertheless an actor's intentional decision to act strategically to take advantage of (rather than to be merely swept along by) impersonal historical forces is the primary determinant of how political outcomes are determined.[68]

By way of pre-empting any objections to the use of the rational choice institutionalist approach, two caveats are in order. First, my research is not meant to be an empirical application of rational choice institutionalism to the privatization of the Russian oil industry. While rational choice institutionalism serves as the underlying framework to explore business–state interactions, alternative explanations such as historical institutionalism will be considered and included if they enhance the completeness of the explanation. There is no intention here of proposing that rational choice institutionalism is the definitive and only approach to analysing public policymaking with respect to the privatization of Russia's oil industry.[69]

Second, the emphasis of rational choice institutionalism on methodological individualism may not appear to be compatible with exploring the interaction among group-based entities in public policymaking. However, groups are a fixture of politics, given that public policy is rarely produced by a single unitary actor but, rather, results from interaction among various collectives of political actors. Notwithstanding, I will revert to the individual level of analysis whenever it becomes pertinent to state the preferences that an actor brings into the public policy arena. In addition, the most relevant actors in a policy process typically act in the interest, and from the perspective, of larger units rather than themselves. As such, I will treat a limited number of large units as composite actors with relatively cohesive action orientations;[70] in other words, a corporation such as pre-2005 YUKOS is treated as a composite, purposeful and rational actor personified by the group leader, Mikhail Khodorkovskiy. YUKOS's corporate interests hence reflect his personal preferences and vice versa.

Therefore, in order to understand the process and outcome of the privatization of the Russian oil industry, this book will consider the individual preferences of political actors, including leading state officials

and representatives of oil companies, as well as the existing and historical institutional incentives (formal and informal) that structure decision-
making by such actors. It is by usefully combining the rational choice
institutionalist approach with other explanations that we can better
advance our understanding of the policy process in the Russian oil
industry.

In this connection, let us examine the previously noted divergence
between the scope of reforms in the oil and gas industries. Back in 1992,
the 'young reformers' had, in fact, advocated de-monopolization in both
the oil and gas industries, in line with their self-interested goal of creating a market economy in Russia by first depoliticizing state control over
the economy. Indeed, the State Committee for the Management of State
Property (or GKI) with the support of Mintopenergo led by Gaydar's
appointee, Vladimir Lopukhin, proposed the creation of several independent gas-producing entities as joint-stock companies, with the
Gazprom state *kontsern* retaining control over the gas pipeline system
and functioning as a pipeline operator.[71] However, as explained below,
the 'young reformers' ended up accepting a suboptimal policy in the case
of gas because of several considerations peculiar to the industry.

One of the structural preconditions concerned the fact that the customer base of the gas industry is much more restricted than oil.[72] Gas is
transported predominantly through pipelines to end-users for residen-
tial and industrial purposes. In contrast, customers for the oil industry
include refineries – which differ in their capacities to turn out various
grades of refined oil – as well as transportation companies, since oil may
be moved via sea, rail or pipelines. There is, therefore, an inbuilt tendency towards a single, all-encompassing structure in the gas industry
as opposed to diversity in the oil sector.

Another structural given is that the three supergiant gas fields in
western Siberia, which in 1991 accounted for over 90% of the country's
gas production, are located in close proximity.[73] Comparatively, the
three largest oil enterprises in 1991 accounted for only one-third of
Russia's total output and were situated further away from one another
than the supergiant gas fields.[74] The geological and geographical features of the gas industry are thus more concentrated than in the oil
sector, lending weight to the argument that Gazprom be retained as a
single, cohesive entity.[75] This tendency to consider Gazprom a natural
monopoly was, in fact, in line with prevailing international practices
in sectors such as gas, electricity, telephony and water. It was only after
the mid-1990s that economists and policymakers began to question
the relevance of natural monopolies, and since then, some countries

have deregulated or privatized formerly state-owned entities in these sectors.[76]

The legacies of the rules and norms of the Soviet economic system were also taken into account by the 'young reformers'. They realized that the power of the general managers of oil and gas enterprises differed significantly because the Soviet oil industry had been answerable to four different ministries, whereas gas had been under the exclusive jurisdiction of the Ministry of Gas.[77] These so-called oil generals were more intensely involved in running the enterprises, since they had to balance the sometimes conflicting interests of the various ministries.[78] As a result, the 'oil generals' came to view their enterprises as personal fiefdoms; this implied that the 'oil generals' would be receptive to incentives aimed at inducing cooperation, particularly if these legalized their *de facto* control over oil companies. Decree 1403 was hence devised as an institution to reward these general managers with gains from cooperating with the 'young reformers' on corporatization and privatization in the oil industry.

The 'young reformers' were also cognizant of historical trends in energy production and use. One of these trends was that unlike gas, there was a severe malaise in the oil industry: between 1980 and 1989, gas production increased by nearly 80% whereas oil output rose by a mere 1%.[79] Indeed, by the mid-1980s, gas had replaced oil as the dominant source of domestic energy; in 1992, gas accounted for 46% of total primary energy supply in Russia, compared to a 30% share for oil and 17% for coal.[80] Also, Russia had plentiful supplies of gas reserves – enough to last for 80 years – whereas its oil reserves were expected to be exhausted within 20 years.[81] The 'young reformers' were also aware that there was already an established clientele and marketing channel for oil exports, which had been a major source of hard currency for the Soviet Union.

Together, these trends implied that the de-monopolization, corporatization and privatization of the oil industry presented a clear-cut, win-win situation for the 'young reformers', whereby the revenue-generating ability of the oil industry could be harnessed with little adverse impact on its role in the domestic economy, where gas was the fuel of choice. In this connection, the 'young reformers' imposed an ever-increasing number of taxes, duties and levies on the oil industry at all levels of government; by 1993, the 40 or so types of oil-related payments laid claim to more than 100% of a company's production profits![82]

Apart from these structural and historical legacies, the situational context of reforms between 1991 and 1992 also shaped the parameters of decision-making by the 'young reformers'. Criminality abounded in exports of gas and, particularly, oil, shortage of basic necessities were *de rigueur*, Soviet and Russian institutions and politicians were locked in a power struggle ostensibly over economic reforms, the Soviet republics and ethnic minorities were restive, and there were fears that Russia itself would disintegrate. Viktor Chernomyrdin, Russia's former prime minister, noted that the government's task at that time was 'to make sure Russia did not come apart. What other vital interests could there be? Russia's existence itself was at stake'.[83] Working under conditions of extreme uncertainty, constant change and time constraints meant that the 'young reformers' did not have the luxury of evaluating all possible options for reforming the oil and gas industry and then selecting the most optimal solution. Instead, they chose a 'satisficing' strategy that met some minimal level of progress towards depoliticization of the economy.

Accordingly, as a first step, de-monopolization and privatization would progress much deeper and faster in the oil sector. It was only in a second, later, phase that the gas industry would follow suit. There is evidence that GKI did, in fact, intend to radically reform Gazprom shortly thereafter. The Anti-Monopoly Committee, which was closely allied with GKI, warned that the presidential decree of November 1992 that created Gazprom contradicted the state programme for privatization, which forbade the creation of a joint-stock company on the basis of an intact and unified *kontsern*.[84] Then in late 1993, GKI prepared a draft decree which envisioned a plan to dismantle Gazprom's monopoly. However, this was never signed into law by Yel'tsin.[85]

From the above example, it should be clear that, on the one hand, while historical institutions and conditions constrained the agenda of the 'young reformers' with regard to reforms in the energy sector, on the other hand, they provided an opportunity to induce cooperative behaviour from the 'oil generals'. From this perspective, accusations that the 'young reformers' regarded policymaking as *tabula rasa*, and that they implemented economic reforms inimical to Russia's historical traditions, appear to be misplaced.[86] On the contrary, Gaydar and his team recognized and harnessed the legacies of the oil industry and its leaders in favour of corporatization and privatization.

Organization of the book

This book attempts to understand the intricacies of the policymaking process in Russia with regard to privatizing the major oil companies. The focus is on the process itself and, therefore, all normative judgements about the fairness, desirability or morality of personal motivations and policy outcomes will have no place here. The main thrust of the argument is that the perception that post-Soviet tycoons have privatized and 'captured' the Russian state, thereby determining the design and implementation of the policies to privatize the oil industry between 1992 and 2006, is an erroneous one. It is too simple and fails to take into account the contribution by Soviet-era general managers of oil companies in creating vertically integrated oil companies and in taking the first steps towards privatization. After all, as Chapters 3 and 4 will indicate, respectively, 15% of YUKOS had already been divested from state-ownership before the loans-for-shares auctions in 1995. Similarly, 8% of Slavneft' had already been sold to private shareholders prior to the first major attempt at privatizing the company in 1997. The said perception also underestimates the extent to which the presidency and other state agencies were able to use oil privatization to achieve their own, separate, interests. Moreover, it ignores the fact that the political power of business can and does vary during different time periods, and that political resources are not synonymous with political power but must be used efficaciously if they are to be converted into power.

Chapter 2 presents an overview of the initiation and spread of privatization-related reforms in the Russian oil industry. The aim is to consider the extent to which external, impersonal factors – including the atmosphere of crisis in Russia in 1991–1992, the 'Washington consensus' and Soviet-era attempts at reforming the oil industry – determined the inevitability of restructuring and privatization reforms in the oil industry. It also examines the role, motivations and behaviour of both state and non-state domestic actors, with a view to identifying the driving force behind the initiation and spread of privatization. The changes which have occurred within the oil industry since 2003 as a result of the YUKOS affair, which some observers have characterized as de-privatization, will be analysed separately in the interests of organizational clarity.

The next three chapters comprise case studies of YUKOS, Slavneft' and Rosneft'. Prior to the YUKOS affair, the first two oil companies were privatized, whereas Rosneft' remained a fully state-owned entity. The choice of the three case studies is a deliberate attempt to minimize any selection bias which may affect our conclusion.[87] While YUKOS and Slavneft' clearly differed from Rosneft' in terms of ownership, and although the method and timing of privatization of YUKOS and Slavneft' (and later the partial privatization of Rosneft') were distinct, the fate of all three companies was determined largely by the same variable in the policy-making process, namely, the dynamics of the triangular relationship between tycoons from the private sector, the *nomenklatura* managers of the oil company in question and state agencies.

Chapter 3 opens with an inquiry into the factors that motivated the 'oil general' Sergey Muravlenko, to create and then sell YUKOS. The chapter suggests that his doubts about giving up his partial ownership and control rights were not translated into action largely because of the gains to be had from cooperating with the sale. In order to ensure a successful divestment, Mikhail Khodorkovskiy and his supporters in the state bureaucracy had to exercise repeated choices in favour of privatization. The chapter continues with an analysis into the YUKOS affair of 2003–2004 and, in particular, why both Khodorkovskiy and Putin declined to uphold the implicit rules of the game that guaranteed secure property rights in return for not challenging the president's authority and power.

Chapter 4 highlights the transient nature of personal connections. These facilitated the rise of Anatoliy Fomin and Mikhail Gutseriev as successive presidents of Slavneft', but were not able to prevent their subsequent dismissals. The fact that the privatization of Slavneft' in 2002 was preceded by failed attempts at divestment in 1998 provides a good opportunity to explore the changes that took place in the intervening period. These relate especially to the shifting dynamics and interest perception among actors within the triangular relationship.

The failure to divest of Rosneft' in 1998, which is often attributed to poor financial conditions and oil prices in 1998, is the initial subject of Chapter 5. It examines the validity of this assumption by reviewing the extent to which this failure was inevitable as a result of the historical legacy of Rosneft' as a *de facto* national oil company and of the wider struggle between rival companies and their political allies, which accompanied the attempted privatisation of Rosneft'. The final part of the chapter then looks at the divestment of 14% of Rosneft' in June 2006 and suggests that the notion of 'state interests' is being manipulated for personal self-interest.

Finally, Chapter 6 draws together the main strands of the arguments in this book concerning the business–state relationship with regard to oil privatization as well as the importance of the often-ignored *nomenklatura* managers in this respect. In particular, the chapter assesses the utility of the rational choice institutional approach in the current research, most notably its disregard for historical influences in the decision-making process.

2
The Initiation and Spread of Privatization

Privatization may be broadly defined as 'the shifting of function, either in whole or in part, from the public to the private sector', such that there is 'increased reliance on private actors and market forces'.[1] The range of activities associated with privatization include divestment of state assets, contracting out public services to private companies and deregulation of monopolies.[2] For the purpose of the current research, I will focus only on one particular aspect of privatization, that is, the process of deliberate transfer and sale, in whole or in part, by a government of state-owned enterprises or assets to private economic agents.[3] This process also includes preparations for the actual sale, such as corporatization, appointing suitable general managers and drawing up the terms of the sale. Given that privatization, and more broadly economic reforms, arouses considerable opposition and win few immediate supporters for any government, why then did the nascent Russian government embrace and perpetuate such a radical policy change in the oil industry?

Corporatization of oil companies

The oil industry that Russia inherited from the Soviet Union was in a state of decline.[4] Oil production had peaked at 569 million tonnes in 1988 and had fallen by 20% at the end of 1991.[5] A leading Russian newspaper even predicted that declining oil production would fail to meet domestic needs, so that by 1992 oil exports would be banned, and Russia would be forced to import oil.[6] The problems faced by the oil industry included low returns on investment, runaway production costs, inefficient use of manpower, maturity of oil wells and low levels of technical innovation. For instance, the capital investment required per tonne of additional oil capacity doubled between 1970 and 1985, which then increased twofold again between 1985 and 1990.[7] Worse, the availability of cheap oil and other energy resources encouraged wasteful energy consumption: Russia's energy-intensity use in 1993 was about five times higher than the average level in the member countries of the Organisation for Economic Co-operation and Development.[8] In fact, the energy sector as a whole consumed the lion's share of investment at the expense of other sectors of the economy[9], thereby creating a dangerously imbalanced industrial policy and resulting in the downturn and stagnation of Soviet economic growth from 7.2% to 5% to 2.3% during the 1960s, 1970s and 1980s, respectively.[10]

Recognizing that changes were required to rejuvenate the Soviet economy, Mikhail Gorbachev, general secretary of the Communist Party and later president of the USSR, introduced his reform programme known as *perestroyka* (restructuring). This affected the oil industry in various ways. First, economic legislation was introduced during the late 1980s to encourage devolution of decision-making from ministries to firms, as well as to kick-start entrepreneurial activities and the formation of a range of economic entities by workers and management, short of fully privatized enterprises. Examples of such legislation included the Law on the State Enterprise (1987), Law on Co-operatives (1988) and Law on Enterprises and Entrepreneurship (1990).[11] As a result, the leaders of state-owned oil production and refining entities obtained decision-making prerogatives over production, finance and marketing at the expense of central ministries. Describing the relationship that had existed prior to these changes, Alekperov noted that as oil production associations, 'we were merely given funds for equipment, and our main goal was to get out more foreign and Russian equipment in order to develop new oil fields. We had no export or import policies of our own, independent of the global politics of the state'.[12] Henceforth, oil entities were given a free hand in decision-making as long as they fulfilled state orders and targets.

Second, between 1989 and 1991, as part of Gorbachev's 'anti-ministerial' campaign, ministries were abolished in favour of *kontserny* and holding companies. This move was based on the perception that industrial ministries had become more powerful than either the Communist Party or its leaders, since they straddled both political and economic functions.[13] In this regard, Gorbachev attempted to reduce ministerial power by transferring the latter's political power to new, representative institutions that he felt he could control, and by redesignating them as entities with commercial purposes. These business *kontserny* would be responsible for covering their costs from their revenues and be given greater powers in determining prices and salaries, while remaining as state-owned entities.[14]

The industrial ministries were initially concerned about their new status, but they quickly embraced it when they realized that the 'gains to be made from entry into a market [were] so heavily skewed in their favour as a result of their inherited power'.[15] The first such ministerial *kontsern* was created in 1989 from the Ministry of Gas by its former minister Chernomyrdin. While its function was indistinguishable from that of a ministry, it was, legally, an independent business entity. Other similar *kontserny* were subsequently created, but when the Oil Ministry tried to do likewise in 1991, it was overruled by the prime minister, Ivan Silayev, who felt that this would result in 'an oil monster which would try to influence oil policies'.[16] Such a ministerial oil *kontsern* would in turn further undermine the authority of his government, which was already weakened by the creation of Chernomyrdin's Gazprom. Silayev therefore agreed only to reorganize the Oil Ministry into a voluntary association known as Rosneftegaz in late 1991 and not as a powerful *kontsern*.

The significance of these changes was as follows. First, they 'gave enterprise directors their first taste of real autonomy'[17] and provided legal grounds for their *de facto* ownership over enterprises that they managed. Second, the proliferation of ministerial holding companies underlined the weakness of central authority vis-à-vis well-connected officials and created a precedent that could be emulated by general managers of companies within the energy industry.[18] Indeed, 'the fact that Gazprom was created as a *kontsern* made them [the management of LUKoil] think of creating LUKoil using the same format and calling it a *kontsern*, since such a term was already used in legislation at that time'.[19] The general managers of three hitherto separate oil production entities therefore began negotiations to merge their enterprises into a single *kontsern* called 'LangepasUrayKogalymneft'' led by Alekperov.

In 1991 Alekperov was appointed as first deputy minister of the USSR Ministry of Oil and Gas. He visited the headquarters of several large Western oil companies and became well acquainted with the vertically integrated structure of operations, whereby a company was involved in all aspects of the oil supply chain ranging from extraction to refining to retailing. Although this had been the norm for major oil companies in the West, it was a novelty in the Soviet Union, where the supply chain was horizontally organized among different oil enterprises. Alekperov therefore decided to emulate the successful organizational structure of

Western oil companies and expanded the LangepasUrayKogalymneft' *kontsern* to include oil refineries and distributors. The legal status of this enlarged *kontsern*, renamed as LUKoil, was officially confirmed by the USSR Council of Ministers Resolution 18 issued in November 1991.[20]

Third, Alekperov's example, the parallel existence of Soviet and Russian ministries overseeing the oil sector after the summer of 1991, and the obvious weakness of Rosneftegaz, further emboldened the general managers of oil enterprises to claim control over the assets that they had managed on behalf of the Soviet state in a chaotic process known as 'spontaneous privatisation'. None of them managed to create additional oil *kontserny* prior to the end of the Soviet Union, but many succeeded in taking advantage of the disarray to sell or transfer productive assets belonging to the state for personal gain. They also managed to make money from selling domestic oil abroad through the use of tax-free export quotas administered by 'friendly' state officials. Oil was also sold abroad through illegal smuggling networks, some of which were abetted or even run by senior state security officers.[21] Given that at the end of 1991, one tonne of crude oil was priced at US\$0.50 in the Soviet Union but sold at US\$100 in world markets, exporters and traders benefited enormously from the price differential.[22] Indeed, between 1992 and 1995, capital flight from the Russian oil industry alone was estimated at US\$7.4 billion.[23]

The crisis in the oil industry at the end of 1991, however, was not an automatic trigger for the initiation of radical policies aimed at divesting state-owned enterprises.[24] The Soviet oil industry was already in crisis during the 1980s. Yet this had not prompted the Soviet leaders to introduce significant changes despite their awareness of the challenges facing the industry. For example, to underline concerns about production problems in Western Siberia, the Soviet Oil Minister at that time, Nikolay Mal'tsev, typically dismissed a number of oilmen during his visits to Tyumen' during the late 1970s and early 1980s.[25] The declining growth in oil production likewise prompted General Secretary Yuriy Andropov, in 1983, to approve an energy programme that recommended energy conservation measures.[26] As for Gorbachev, he acknowledged that continuing oil-production shortfalls in Tyumen' were 'creating difficulties in the agricultural sector', and that the problems in the oil industry were 'not decreasing, but increasing'.[27] In short, Soviet leaders did not feel that the problems in the oil industry directly imperiled their tenure as leaders and merely shifted the distribution of resources instead of undertaking fundamental reforms. Hence, the book argues, the crisis hypothesis by itself cannot adequately explain the decision to corporatize and privatize the oil industry. This is because 'some dissatisfaction with the economic status quo is a necessary, but not sufficient, condition for the launching of reform efforts'.[28]

Another variable that is purported to have shaped the decision to change the structure of ownership in the oil industry relates to the spread and legitimacy of privatization as a public policy, both globally and in Russia, prior to the initiation of reforms in late 1992.[29] At the global level, the post-war growth of state control in the economy was challenged by the launch of large-scale privatization by Margaret Thatcher's government in the United Kingdom in the 1980s.[30] Since then, privatization has been adopted as a core economic and political aim by many countries in western Europe, Latin America and Asia. According to John Nellis, a renowned expert on privatization, 'as many as 100,000 firms and business units formerly owned and operated by governments, in every industrial, commercial and service sector, in every geographical region of the world' have embraced some form of privatization, while only a handful of countries have shied away from it.[31] Privatization has resulted in a dramatic reduction in state-ownership from a high rate of 8.5% of GDP in developed countries and 16% of GDP in developing countries in the early 1980s to less than 5% of GDP today.[32] Within the former communist countries of Europe, around US$100 billion worth of privatization receipts have been generated between 1990 and 2003,[33] although this figure underestimates the scope of privatization since a large number of state-owned enterprises there were divested free of charge.

Privatization was also making inroads within the oil industry in the 1980s and early 1990s as exemplified by the sale of hitherto state-owned companies such as British Petroleum, Enterprise Oil and Britoil (United Kingdom), Elf Aquitaine (France) and Repsol (Spain).[34] Nevertheless at the end of 1991, when political elites in Russia were engaged in a debate about reforms, world oil production was still dominated by countries, most of them in the Middle East, with a predominantly state-owned oil industry. Indeed, major oil producing countries accounted for 77.3% of world oil production in 1991, of which only one-quarter was produced in countries with privately owned oil companies (namely, the United States, Canada and the United Kingdom).[35]

In Russia, the global spread and popularity of privatization were reflected in the debates on economic reform between 1988 and 1991. Although the competing proposals differed on whether reform should take place within the socialist system or through the adoption of market-oriented principles, for the most part 'the object of the debate was not the correctness of the process itself but which were the most effective and socially acceptable methods of privatization'.[36] This near-consensus was the result of collective learning by elite groups over time and was, in particular, their response to past experience with the inadequacy of economic reforms within the Soviet, socialist system.[37] In this respect, privatization as an idea promoted by epistemic communities was a good 'fit' with the existing elite consensus on the need to give more rein to private enterprise by reducing the scope of state-ownership in the economy.

Just like the crisis in the oil industry, however, the acceptability of privatization globally and within Russia did not trigger an automatic withdrawal of the state from the economy. On the contrary, the defence, telecommunications and natural resources sectors were among the 30% of state property excluded from the mass privatization programme.[38] Also, the consensus on the principle of privatization soon gave way to differences on implementation, particularly with respect to its actual form and scope. Moreover, as will be discussed shortly, the privatization programme that was specially designed for the oil industry gave the state stakes of between 45 and 51% in the newly established oil companies. The most significant contribution of the idea of privatization was to reduce the range of alternatives available to decision-makers in undertaking economic reform, but ultimately, it fell to political and economic elites to engage in the 'art of the possible' and make privatization workable in the oil industry.

Rather than notions of the 'crisis hypothesis' or 'policy bandwagoning', the primary variable explaining the initiation of reforms in the oil industry was the conscious choice made by elite actors within the state and non-state sectors.[39] They were the ones who interpreted the situation in the oil sector as a crisis that demanded more far-reaching changes than had previously been attempted by the Soviet leadership. Moreover, they also chose to make use of the spread and acceptability of privatization globally and in Russia to justify economic changes that chipped away at the hitherto sacrosanct notions of state-ownership of the economy. Vladimir Mau, a well-known Russian economist, asserts that the 'young reformers' used the International Monetary Fund (IMF) as a convenient stalking horse to force the pace of reforms. In this respect, 'a good part of the "IMF conditions" were developed in Moscow, not in Washington. Russian politicians [were] the ones who initiated many of these conditions. And then only were they "imposed from outside". This is a typical way for a weak government to launch unpopular reforms'.[40]

Alarmed by the state's rapid loss of control over the vital oil industry, Gaydar announced his determination to put an end to spontaneous privatizations in the oil industry, which he considered as 'one of the most criminalised spheres of administration'.[41] Towards this end, Gaydar and Chubays proposed immediate de-monopolization, liberalization of prices and exports, privatization and competition within the energy sector. For the oil industry, this implied that the existing 300 oil entities, ranging from production, refining to marketing companies, would be included in the mass privatization programme as the fastest way to remove state-ownership over a troubled sector. This would have meant the break-up of dominant, large oil associations and the appearance of hundreds of small, privately owned and independently functioning oil units. At that time, Gaydar and Chubays opposed proposals by Lopukhin and the 'oil generals' to restructure the industry prior to privatization, so as to create several large, vertically integrated companies that would be better equipped to compete on world markets.[42] For Gaydar and Chubays, such proposals appeared to be an attempt to delay the progress of privatization and, as such, were inimical to their preferences for market-oriented reforms.[43]

Apart from Mintopenergo, the Gaydar/Chubays plan for the oil industry was also challenged by regional elites in oil-rich areas, the *nomenklatura* managers of oil enterprises as well as the agro-industrial ministries and sectors of the economy, for reasons that will be explained later. In May 1992, the ranks of liberal reformers were weakened by a cabinet reshuffle aimed at placating the agro-industrial sector. Yel'tsin dismissed Lopukhin as head of Mintopenergo and appointed three members of the agro-industrial lobby to government positions, including Chernomyrdin, who was named the deputy prime minister in charge of the fuel and energy sector.[44]

Under the patronage of Chernomyrdin, the preferences of the 'oil generals' and agro-industrial sector prevailed over the Gaydar/Chubays plan. In August 1992, the oil and gas industries were exempted by presidential decree from the general privatization programme. In September 1992, a conceptual outline to reverse the deep crisis in Russia's energy policy was presented and approved by the government. It called for pipelines to remain under the control of the state, recommended the acceleration of the corporatization of oil companies and noted the progress made in studying the idea of creating 12 major VIOCs.[45] Then, on 17 November 1992, a separate programme for the privatization of the oil industry was announced via presidential decree 1403.[46] According to Mintopenergo, which was then overseen by Chernomyrdin, this gradual programme for

the 'first stage of the privatisation process' of the oil industry was necessitated by the fact that the industry's 'priority importance to the government didn't permit application of the harsh, fundamental measures
capable of solving its problem'.[47]

Gaydar and Chubays hence appeared to have been defeated by the
new political environment, whereby proponents of gradual change
were in the ascendancy. Nevertheless Gaydar and Chubays may have
accepted, albeit belatedly, that their original plan was unrealistic and
agreed with Mintopenergo that 'it was impossible to build an economy
of small pieces'.[48] According to a GKI insider, this acceptance was the
product of on-the-job learning. In this connection, GKI's task was to
'carry out privatisation while retaining its initial principles, but without
affecting manageability and without harming the economy by introducing changes recklessly ... This task was not voiced and fully realised
in the very beginning. We started to understand it in the process of
work'.[49] Given that these 'young reformers' had kept the bigger, strategic goal in mind – reducing the scope of state-ownership – in acquiesc-
ing to the different pace of privatization in the oil and gas sectors (as
explained in Chapter 1), it is arguable that they perceived of decree
1403 in the same light. Hence their decision to cooperate on, rather
than to 'defect' from, creating and instituting new rules of the political
game.

Decree 1403 contained several key provisions. First, oil production,
refining and marketing enterprises were corporatized. This provision
was aimed at reaffirming formal state-ownership of the oil industry,
thus putting an end to the 'spontaneous privatizations' of the previous
years. Secondly, the decree announced that LUKoil, Surgutneftegaz and
YUKOS would be established as VIOCs. However, given that at this
stage 'management's instincts [were] still rudimentary',[50] the state
would retain a controlling share in these VIOCs (amounting to 45% of
the shares in the parent or holding company) for three years. These
shares would be jointly managed by GKI, Mintopenergo and the Anti-
Monopoly Committee. As for the remaining shares in the parent company and those in its subsidiaries, they would be divested among
employees, the general public and investors. Third, the Soviet-era oil
concern, Rosneftegaz, was abolished in place of Rosneft'. This state-
owned enterprise was tasked to manage in trust the shares of oil enterprises that were not part of the above-named VIOCs. Finally, the limit
on foreign share-ownership in VIOCs was 15%.

In compliance with the decree, resolutions were issued by the Russian
government to create VIOCs. Surgutneftegaz was formally incorporated

through a government resolution dated 19 March 1993, LUKoil through government resolution 229 of 5 April 1993[51] and YUKOS through resolution 354 of 15 April 1993.[52] In May 1994, another four VIOCs were created by resolution 452, namely, Siberia Far East Oil Company (SIDANKO), Slavneft', Eastern Oil Company (VNK) and ONAKO. They were followed by presidential decree 327 on 1 April 1995, which established Rosneft' as a VIOC and empowered it to represent the state's interest in production-sharing agreements in Russia (that is, to negotiate on the state's behalf with foreign oil companies and to assume the state's share of production stemming from such agreements).[53] The Tyumen' Oil Company (TNK) was created as a VIOC by the same presidential decree of 1 April 1995 and sanctioned by the corresponding government resolution 802 of 9 September 1995. Finally, presidential decree 872 of 24 August 1995 ordered the creation of the last VIOC, Sibneft', and the government duly obliged with an appropriate resolution issued on 29 September 1995.[54] VIOCs therefore became the key organizations in the oil industry, and 10–15 such companies were expected to be created in the course of privatization, as reflected in Mintopenergo's 'Concept for the Management of the Country's Oil Complex', which was approved in September 1994.[55]

Policy Initiation: Actors and their motivations

From the above account, it is clear that the key actors who initiated the process of ownership change in the oil industry were GKI, Mintopenergo and influential 'oil generals'. Viktor Ott, head of the Oil Committee of Mintopenergo that drafted the 1403 decree, noted that all relevant issues were discussed with these managers 'because otherwise [the decree] could not progress any further ... it was collective labour'.[56] What were the motives and calculations that guided the various actors in making decisions concerning the corporatization of the Russian oil industry?

According to public choice theorists, the head of a bureaucratic agency is assumed to be a 'budget-mazimizer', whose actions are the result of his interest to increase the amount of budgetary funds allocated to his department.[57] This is because he is motivated by one or more of the following considerations that accrue with an ever-increasing budget: increasing his own remuneration or on-the-job perks, savouring the higher prestige that comes from managing a major department with a growing number of staff and widening scope of responsibilities, or being able to implement public and social programmes that he cares about.

The assumption of a budget-maximizing bureaucrat – regardless of whether his self-interest is based on selfish or altruistic reasons – implies that he will always engage in behaviour aimed at supporting big government and state intervention in the economy. This is because the administration of regulations and quotas and other forms of 'red-tape' typically creates opportunities for bribes, rents or larger budgets that will increase the funds available to his department. In this connection, 'the best way to limit rent-seeking is to limit government'.[58] In Russia, this formed the basis for the determination of Gaydar's team to 'depoliticise economic life'[59] by reducing political control over prices and enterprises.

Mintopenergo, particularly after mid-1992, was the classic embodiment of a budget-maximizing bureaucratic agency in favour of a larger role for the state. It was an unwavering supporter of the use of production-sharing agreements because these were expected to enhance the scope of the Ministry's responsibilities in the oil industry.[60] With regard to privatization, Mintopenergo consistently opposed attempts by GKI to reduce the extent of state-ownership over oil companies. According to Yuriy Shafranik, head of Mintopenergo between 1993 and 1996, when VIOCs were being created, '51% was left for the state but we intended to leave more … it was a compromise with Gaydar's group',[61] which wanted to privatize everything. Shafranik was well aware that the continued transfer of state-owned assets to the private sector would imperil the relevance of his ministry and, by extension, his own position in managing the oil industry. In this connection, he objected to the participation of oil companies in the loans-for-shares scheme[62] and instead proposed the creation of a national oil company with Rosneft' as its major component. It is noteworthy that during the first few years of Mintopenergo's existence, suggestions were, in fact, made for the ministry to be dissolved in view of the growing number of corporatized VIOCs and the ministry's seeming inability to reverse the decline in oil production. For example, one of the 'oil generals' claimed that it was 'necessary to dissolve Mintopenergo as soon as possible' because its specialists were not well-qualified to tackle the tasks at hand.[63] Faced with the constant threat of dismissal, Shafranik was probably 'building himself an emergency landing strip', namely, to be named head of the proposed national oil company.[64]

The above example of Mintopenergo notwithstanding, sometimes bureaucratic agencies may choose to reject the budget-maximizing maxim and instead, support a reduction in the state's role in the economy. Specifically, the promise of pecuniary benefits (that is, material rewards and high-paying jobs with attractive perks) may prompt bureaucrats to promote the expansion of market-oriented policies. Steven Solnick, Stephen Whitefield, Simon Johnson and Heidi Kroll have noted, for example, that the Soviet *nomenklatura* were in favour of decentralizing state control in the economy in the late 1980s because they were in a position to benefit materially from such changes.[65] Consequently, they 'exchanged "Das Capital" for capital'.[66] Moreover, bureaucrats who have proven themselves to be capable in implementing pro-market policies, or who have used their positions to serve the interests of private firms, have gained lucrative jobs in the private sector. Examples include Pyotr Aven, who joined Al'fa bank after he resigned as Minister for Foreign Economic Relations under Gaydar, and Aleksandr Samusev, who joined YUKOS following a career at Mintopenergo where he helped to draft decree 1403 and at the Ministry of Finance where he allegedly favoured Menatep's bid for YUKOS.[67]

Another reason why bureaucrats may sometimes support policies that favour a reduction in the state's role in the economy relates to bureaucratic competition over turf matters. Timothy Frye has argued persuasively that in Russia, the Federal Commission initiated the creation of a market-supporting institution, the broker's association known as NAUFOR, in order to prevent incursions by other agencies, including the Central Bank and Ministry of Finance, into its jurisdiction over the corporate equities market.[68] Likewise, Mintopenergo under Lopukhin supported the creation of VIOCs partly as a tool with which to wrest control over oil refineries away from its rival, the Ministry of Industry.[69] In fact, during this time, Mintopenergo was the lead agency with respect to restructuring the oil industry; some of its experts drafted decree 1403 while others worked with American consultants on a new oil and gas law for Russia.[70] It started favouring a larger role for the state only after Lopukhin's departure and as a defensive response to GKI's growing role in the oil industry as a result of privatization, which resulted in bureaucratic competition between the two agencies.

In the case of GKI, it behaved contrary to the conventional wisdom that bureaucrats will always seek to expand the role of the state. One way to explain why GKI eschewed this traditional bureaucratic goal in favour of limiting the role of the state relates to the behaviour of political incumbents. The latter are generally assumed to value political longevity above all goals since they cannot pursue their pet projects if they are not in office; in this regard, they are interested to retain political office for as long as possible. Accordingly, a rational, tenure-maximizing political incumbent may be motivated to delegate part of his policy-making authority to a team of technocrats, who are empowered to introduce reforms that favour his political constituency, which in turn will vote for him.[71]

In Yel'tsin's case, he turned to the 'young reformers' for various reasons. First, he calculated that supporting Gaydar's team and their programme of 'shock therapy' was the fastest way to consolidate his own power and undermine rivals from the Soviet leadership.[72] Aven the former Minister, noted that Yel'tsin was 'interested only in power', and that he therefore needed 'a team that would be very aggressive in throwing out all the old bureaucrats'.[73] Yel'tsin understood that his anti-Soviet and pro-reform stance was popular with the electorate, and that Gaydar's team of 'young people without any hang-ups, independent thinkers raring to go, . . . arrogant young upstarts'[74] shared a similar passion to destroy all vestiges of communism in Russia.

Moreover, the use of 'shock therapy' to create a market economy in Russia was consistent with Yel'tsin's own experience of the failures of the Soviet economy, as well as his own style of management, which tended towards the need for struggle, campaigns, mobilization and target storming.[75] Indeed, Yel'tsin has explained that he chose 'shock therapy' over a more gradualist programme of economic reform because he 'couldn't force the people to wait once again, to drag out the main events and processes for years. If our minds were made up, we had to get going!'.[76] He subsequently decreed the creation of GKI and enacted laws that gave it a major role in matters of privatization across all industries, including oil, much to the chagrin of Mintopenergo.[77]

The above discussion suggests that one possible reason why Gaydar, Chubays and the successive heads of GKI supported a reduction in the state's role in the economy was because of Yel'tsin's decision to delegate power over economic affairs to them.[78] They therefore had an incentive to act as 'bureau-shapers' in aligning their interests with those of their patron, Yel'tsin, in favour of liberalization and privatization.[79] In return, they benefited from proximity to the power centre and the high level of discretion over economic policies, despite not having been elected to office. This purely institutionalist explanation as to why the 'young reformers' embraced a non-traditional bureaucratic goal in seeking to reduce (rather than increase) the role of government is, nevertheless, inadequate. It fails to address the issue of why the 'young reformers' wanted to obtain power over economic policymaking in the first place. The answer is that the primary, albeit not the only, motivation of the 'young reformers' was to introduce market-oriented reforms in Russia.[80]

In the case of Chubays, he is said to be driven by this desire, which is akin to a mission. 'Not ambition or concern about career advancement,

even though needless to say, they exist, but they are not the basic motivation for work . . . The sense of mission is the main well-spring of his
behaviour'.[81] More precisely, the crux of a market economy was, for
the 'young reformers', private ownership of property. As such, they
expended much time and energy towards this end, often to the detri-
ment of other aspects of creating a market economy – such as competition, regulation and institution-building – on the assumption that
privatization 'will make the whole economic reform process irreversible
because to reverse economic reform will mean for millions of people to
say farewell to their profit'.[82] This focus on private, rather than state,
ownership of property also partly explains why the 'young reformers'
were willing to acquiesce with insider privatization (as embodied by
option 2 of the mass privatization programme) and with loans-for-shares
since, for them, any private owner was preferable to the state. Indeed,
'just as the Bolsheviks had built communism by transferring the means
of production from private owners to the state, so [Chubays] believed
that his central mission must be to undo that transformation, and
return the property to private owners'.[83]

The creation of a market economy through the establishment of private ownership of property was, therefore, in the rational self-interest
of the 'young reformers', since such an outcome would maximize their
own preferences or utility, particularly if they were the ones in charge
of directing such an economic transformation. In view of Yel'tsin's
self-regarding willingness to delegate his power over economic
policymaking to them, and the fact that the 'young reformers' had no
electoral mandate of their own, they were constrained and obliged
to support Yel'tsin's presidency without question, even when he criticized them.

What of the motivations of the general managers of oil production
and refining associations who played an equally significant role in this
process?[84] Initially, the 'oil generals' considered privatization, particularly the 'shock' variant proposed by Gaydar and Chubays, a threat to
their *de facto* control and ownership of oil enterprises since the most
likely buyers with enough money to pay for these enterprises would be
foreigners. It was this *de facto* control that had bestowed upon the 'oil
generals' their economic clout, authority and status, and which had
allowed them to engage in various transfer-pricing and smuggling
schemes that earned them enormous rents. In addition, they felt that
the liberalization of energy prices and the elimination of oil export quotas would reduce the avenues through which they could further enrich
themselves.[85] However, they quickly threw their support behind the

revised privatization plan outlined in decree 1403 when they realized that for at least the next three years, they would have *de jure* control over their enterprises: they could hence continue to engage in asset-stripping or make preparations for the kind of insider privatization that would be advantageous for themselves.

In this connection, the book argues that the *nomenklatura* managers of oil enterprises were hardly 'followers, rather than leaders, of the radical reform'[86] and questions the accuracy of the state-centred view that the government 'set out to fundamentally restructure the sector . . . by combining groups or existing entities and corporatising them to create vertically integrated oil companies'.[87] In fact, the 'oil generals' triggered a response by the state through their 'spontaneous privatisations' of oil enterprises and attempt to create VIOCs. It was their ambition to retain their own fiefdoms that precluded the formation of the West Siberian Oil and Gas Corporation. It was their self-interest, in some cases together with private sector backers, that dictated the composition and number of VIOCs. For instance, the tycoon Boris Berezovskiy managed to create Sibneft' by co-opting the general manager of oil producer, Noyarbsneftegaz (which wanted to break away from Rosneft'), allegedly ordering the murder of the obstinate general manager of Omsk refinery who refused to go along with the plan, getting the head of the oil trading firm cooperating with Omsk refinery thrown into jail, securing the support of the regional governor by employing the latter's son and finally obtaining the support of influential figures within the government and presidential administration.[88] As for YUKOS and Slavneft', case studies of which will constitute the next two chapters, their general managers faced little significant interference from the state and had enormous leeway in selecting the companies with which they wanted to merge.

The initial opposition of the regional elites to privatization was, in many ways, a means to wrest more power away from the centre.[89] In January 1992, for instance, political elites in oil-rich Tyumen' proposed the creation of the West Siberian Oil and Gas Corporation. Organized along vertically integrated lines, the corporation would control the oil and gas enterprises in the region, decide on how best to use these resources for developing the *oblast'* and serve as an intermediary between these enterprises and the ministries in Moscow.[90] Shafranik, at that time the governor of the Tyumen' *oblast'*, frankly admitted that such 'perspectives about the development of Tyumen' *oblast'* were directly connected with the idea of broadening economic self-sufficiency'.[91] Indeed, as president of the Russian Federation, Yel'tsin encouraged such aspirations in order to gain the support of regional leaders in his battle with Gorbachev and the Soviet state.[92]

As for the agro-industrial sector, the fear was that privatization and liberalization in the oil sector would result in higher oil prices.[93] This would increase production costs and bankrupt many of their enterprises which were already reeling from reforms implemented in January 1992 that had freed 90% of prices in the economy. The enterprise managers worried that they would no longer be able to engage in asset-stripping and self-enrichment and warned that bankruptcies would result in payment arrears, mass unemployment and social unrest.

From the above account of the first phase of privatizing the oil industry, it is clear that the adoption of a separate privatization programme for the oil industry was not simply Yel'tsin's 'attempt to appease the resurgent conservatives and Communists who resent[ed] Russia's turn in the direction of market economics'.[94] Indeed, to avoid courting political controversy, the status quo in the oil industry could have been maintained. That is, Rosneftegaz could have been retained as a corporatized state-owned entity without being broken up into separate companies, as was the case with Gazprom or Rosugol' in the coal industry or Unified Energy Systems in the electricity sector. Rather, the fact that any structural reform of the oil industry was even on the policy agenda was a result of cooperation among key actors from the state and non-state sectors to protect and enhance their self-interests. Gradual privatization presented Mintopenergo with a valuable opportunity to retain bureaucratic oversight of the development of the industry; it allowed the 'oil generals' to legalize their *de facto* control over their enterprises while keeping at bay private sector businessmen in the short term, and it dovetailed with attempts by the 'young reformers' to reduce state intervention in the economy.

Sale of oil companies

As mentioned earlier, corporatization was only the first step in privatizing the oil industry. After three years, the state was supposed to sell-off its shares (amounting to between 45% and 51%) in the holding companies that had been created during the first half of the 1990s. Between the end of 1995 and 1997, the state divested most of these shareholdings in a second phase of privatization known as 'cash privatization', during which a variety of methods was utilized, including loans-for-shares transactions and investment tenders.[95]

The loans-for-shares scheme was first presented to the Russian cabinet in March 1995 by a consortium of Russian banks led by Vladimir Potanin, chairman of ONEKSIMbank. The proposal was modified and approved for implementation by presidential decree 889 'On the procedure of pledging the shares in federal ownership in 1995', issued on 31 August 1995. The mechanics of the scheme provided for the state to auction-off the right to manage its stake in selected state-owned companies in exchange for bank loans. This stage was known as 'trust auctions'. Should the state fail to repay any part of the loan by 1 September 1996, the lending bank would be permitted to sell the shares held in trust through 'collateral auctions', and thereafter divide the profits of the sale between itself (30%) and the government (70%). As a result of the trust auctions, the government received credits of about US$800 million, and when it defaulted on the loan repayment, its collateralized shares in 12 state-owned companies in the metallurgical, maritime shipping and oil industries – including YUKOS, Surgutneftegaz, LUKoil, SIDANKO and Sibneft' – were divested.[96]

Accounts of the motivations behind the loans-for-shares scheme centre on four possible explanations. The first explanation, which was advanced by the 'young reformers' and their bureaucratic agencies, regarded loans-for-shares as an 'unorthodox fund-raising scheme'[97] aimed explicitly at securing revenues to relieve pressure on the federal budget. According to Al'fred Kokh, deputy chairman of GKI, there was a 'perennial need for cash yield for the government ... to justify privatising the Russian economy by showing our countrymen that we could keep turning it upside down and shaking trillions of rubles out of it and into the wide-mouthed basket of the federal budget'.[98] Indeed, according to the federal budget for 1995, 9.1 trillion rubles were to be derived from the proceeds of privatization. However, by August 1995, less than 20% of this amount had been raised.[99] The problem of low privatization receipts was compounded by a series of draft laws passed by the communist-dominated Duma in an attempt to discredit privatization and undermine the reformers ahead of the 1995 Duma elections. These laws prohibited the sale of state-owned blocks of shares in companies producing strategically important products. They included companies engaged in oil extraction, refining and marketing, which was exactly what the GKI had been relying on for 70% of the privatization receipts for 1995.[100]

The loans-for-shares scheme thus offered the liberal reformers a way out of this quagmire, since bank credits could be used to offset the budgetary shortfall. The collateral auctions could then be held later, after the president had exercised his veto against the new laws. Sergey Molozhaviy, the then deputy minister of GKI, confirmed the causal relationship between the Duma's prohibition and the loans-for-shares scheme when he noted that since 'it was not allowed to sell shares in oil companies in 1995, GKI invented the scheme whereby shares were not sold but placed in trust management'.[101] Crucially, exchanging shares as collateral for bank credits was regarded as a non-inflationary way of financing the federal budget since it would not jeopardize the reformers' hard-won battle to reduce hyperinflation. Indeed, the average monthly rate of 16% in 1995 was the first time that inflation fell below 20% since the start of reforms in 1992.[102]

It is noteworthy that as early as mid-1994, the Ministry of Economy had already explored this possibility of obtaining loans from foreign banks by using state shares as collateral. Simultaneously, the Ministry of Finance had also considered taking loans from Western investment companies in return for placing shares of Russian companies on international stock markets. Although neither of these initiatives were pursued further, they indicated that the young reformers were sufficiently worried about financial stability and were prepared to obtain loans backed by state shares.[103]

There are some, though, who cast doubts on the explanation proffered above – that loans-for-shares was aimed at resolving budgetary pressures – because of the relatively low prices fetched at the auctions. One critic noted that 'the asking prices for Surgut and SIDANKO [were] as close to free as you can get when buying an oil company'.[104] Indeed, 40.12% of Surgutneftegaz was exchanged for US$88 million worth of credits to the state, whereas this stake was worth US$340 according to its domestic market capitalization.[105] Similarly, at the loans-for-shares auction in November 1995, the state auctioned 5% of its shares in LUKoil for US$35.01 million of credits. However, according to its capitalization on the Russian stock market, 5% of LUKoil's shares was worth at least US$180 million. The company's value to foreign investors was even higher: in August 1995, it received US$250 million from the American oil company ARCO as payment for 5.6% of its shares, and in March 1996, LUKoil obtained a further US$90 million from ARCO for an additional 2.99% of its shares.[106] In the case of SIDANKO, ONEKSIMbank paid US$130 million in December 1995 to manage in trust the state's 51% share. Two years later, BP paid ONEKSIMbank US$571 million for a mere 10% stake in SIDANKO.[107]

Attempts to explain the low auction prices as a consequence of the need to discount the risk of investing in an as yet unstable Russian economy were not really convincing, since foreign investors were clearly willing to pay much higher prices for shares in Russian oil companies than amounts received at the auctions.[108] Furthermore, conducting loans-for-shares auctions at a time of market uncertainty – investor confidence in the Russian market was still shaky after Black Tuesday on 11 October 1994 when the ruble fell by 27% against the US dollar in a single day – was never going to generate maximum revenues for the state. In any case, the fiscal benefit of privatization under the 'young reformers' was rarely a major motivating factor: privatization accounted for less than 1% of annual budgetary revenues and 0.1% of GDP on average between 1992 and 1999.[109] It is therefore highly likely that loans-for-shares was 'motivated by factors other than profit maximisation: it [was] political'.[110]

The second explanation as to why loans-for-shares was initiated was that it served as a panacea to correct problems with insider ownership. Under the mass privatization programme, almost three-quarters of enterprises voted for the so-called option 2, which gave employees the right to buy 51% of a company's voting shares. Over time, the bulk of the shares of privatized enterprises ended up in the hands of the general managers, who bought up enterprise shares from their employees.[111] The increasing concentration of shares in the hands of 'red directors' was alleged to be the reason for the continuing lacklustre performance of the Russian economy, since such insider-owned enterprises tended to shy away from vital restructuring tasks. In this connection, Potanin claimed that 'during the auctions, it was not possible to declare this ... but the real reason of the auctions was to bring normal management to sizeable companies and to break the Red Directors' lobby. It was the most important thing'.[112]

However, it is questionable whether loans-for-shares was aimed primarily at restructuring enterprises controlled by *nomenklatura* managers. In the first place, the original five-year moratorium on sales of collateralized shares was shortened to just under one year. The reduced moratorium on sales meant that banks would not necessarily have to restructure the enterprises in order to obtain a high margin on returns. This was because the expected inflow of foreign money into Russia after the 1995 auctions would increase the prices of the collateralized shares and tempt banks to sell these shares. ONEKSIMbank's subsequent sale of 10% of SIDANKO to BP appeared to reinforce doubts about the extent to which restructuring was a prime motive of the loans-for-shares scheme.

Moreover, there were alternative means of restructuring enterprises. In this respect, a second banking consortium, known as the Reliability Club, advanced an alternative proposal to restructure state-owned enterprises.[113] The Club proposed to provide credits to the state budget, extend loans to specific state-owned enterprises backed up by state guarantees, and joint participation between the state and the Club to implement investment and restructuring programmes for the enterprises. In return, they expected favourable treatment with regard to the sale of state-owned blocks of shares in corporatized entities, including the oil companies.

However, the Club's proposal was never seriously considered by the liberal reformers even though there were good arguments to allow banks to become owners of enterprises undergoing restructuring, as demonstrated by the German model of capitalism.[114] The fact that most of the members of the Club were Soviet-era banks was objectionable to the liberal reformers since it would have compromised their key motivation: to ensure that the spectre of communism never returned to the country by quickly transferring as much property as possible into the hands of private owners. Political overtones therefore coloured much of the discussions about loans-for-shares.

The third plausible reason for loans-for-shares centred on private, commercial banks and their rational inclination to engage in profit-enhancing activities.[115] Under the 'authorised bank' system of the early 1990s, government ministries used only certain approved commercial banks to deposit or channel government funds, such as credits or wages to state-owned enterprises. In view of hyperinflation between 1992 and 1994, the banks paid the ministries little or no interest on these funds, which were instead used to speculate on foreign currency movements. The use of 'free money' to engage in such speculative activities earned the banks enormous profits of up to 10% of Russia's GDP.[116] In order to give the banks an incentive to redirect their stake in a high- to low inflation environment, the 'young reformers' encouraged banks to earn profits from lending activities instead.[117] In this regard, selected banks were given new avenues through which to make profits, namely, by financing government-borrowing through the acquisition of high-yielding treasury bonds (profits from which were tax-free) and of shares in lucrative companies through the loans-for-shares scheme. For the commercial banks, any rise in inflation henceforth would have reduced the real value of treasury bonds and credits extended through the trust auctions of December 1995, since it would have cost the government less to repay the loans. They 'viewed participation in industry as a logical next step',[118] since currency speculation was no longer as profitable and the 'authorised bank' system was about to be eliminated.

Potanin was clearly driven by a profit-seeking motive to propose loans-for-shares to the government. He not only owned ONEKSIMbank but was also keen to acquire the lucrative state-owned Norilsk Nickel, which produced 20% of the world's nickel and cobalt, 40% of the world's platinum and was Russia's sixth-largest company in 1995 in terms of sales turnover. However, attempts to explain the plan's adoption as a classic example of state capture by predatory big business may be inaccurate, since the flow of benefits also accrued to the state in the form of economic and political returns. These included finally achieving monetary stability, by bringing inflation under control by 1995, and imperatives of coalition management (discussed below). Loans-for-shares was therefore more of an exchange than outright capture. In the first place, GKI managed to modify Potanin's original plan in such a way that the amended plan was more advantageous to the state: instead of serving only budgetary ends, it now met the political and electoral needs of Yel'tsin's administration as well through the two-step process of auctions before and after the presidential election, as explained below.

Moreover, the fact that Potanin's consortium did not succeed in owning many of the enterprises they were interested in, suggests a limit to the focus on state capture as the main explanatory variable in the adoption of loans-for-shares. Between August and September 1995, the number of state-owned enterprises whose shares were to be collateralized under the loans-for-shares scheme was reduced from 136 (as proposed by the banking consortium) to 29, and in the end, the state-owned blocks of shares in only 12 enterprises were successfully auctioned off. One member of the consortium opined that it was 'annoying, of course, that the final list [was] shorter than was assumed initially'.[119] For example, LUKoil, the largest and most valuable company in the list, managed to reduce the portion of state-owned shares to be included in the auction from 15% to 5%.[120] In fact, of the five oil VIOCs that were included in the auctions, LUKoil, Surgutneftegaz and Sibneft' successfully fended off the advances of the banking consortium by using insider, affiliated or partner companies to secure the winning bids on behalf of the management of these companies. It would, hence, be a mistake to view loans-for-shares as solely the result of the banks' political muscle because such sell-offs 'generally took place when the banks had managed to forge alliance with key industrial companies: hostile acquisitions were extremely rare and were fiercely contested'.[121] As for YUKOS, the fact that Menatep bank won the first and second auctions and ended up with 78% of YUKOS's shares, was not due solely to Menatep's ability to impose its will on the state. Instead, it was the result of a mutually beneficial alliance that had been worked out over the last two years between the management of Menatep and YUKOS, as the next chapter will explain.

This brings the discussion to the fourth, and probably most convincing, reason for the loans-for-shares scheme: that it was a tactic to manage coalition partners since in 1995, the future of Yel'tsin's administration and of privatization appeared to be threatened by a communist victory at the polls. Indeed, for the first time since polling began in 1988, more Russians opposed economic reform than favoured its continuation, according to a survey conducted in January 1995.[122] By August, only 16% of Muscovites polled approved of Yel'tsin's performance as president, down from 55% in 1991, despite the fact that Moscow was a city traditionally favourably disposed towards him.[123]

For the bankers who proposed the loans-for-shares plan, the growing power of the communists threatened the security of their property rights. Indeed, a leading member of the Communist Party was quoted as saying that 'one of the options . . . is to nationalise about 200 of the largest enterprises . . . With respect to commercial banks, we have to be firm since they pose a threat to all of society'.[124] For Yel'tsin and the GKI, the Communist Party led by Gennadiy Zyuganov undermined Yel'tsin's political longevity and the consolidation of market-oriented reforms, including privatization. Yel'tsin and the GKI were thus inextricably linked since Yel'tsin's declining approval ratings were 'not just Yel'tsin's problems. This negative assessment of the President's actions extended to democrats in general'.[125] Therefore, loans-for-shares was a 'political pact . . . to ensure that Zyuganov did not come to the Kremlin'.[126]

To modify what was originally an economic lifeline for the state and a profitable plan for the banks into a political pact, the 'young reformers' designed a two-step process whereby the banks would initially be given trust management of shares in valuable state-owned companies in late 1995 in return for credits. The final, formal transfer of ownership of these companies to the banks would only take place in September 1996 – that is, after the presidential elections in June 1996 – and only if Yel'tsin emerged victorious. In a further amendment to Potanin's plan, the state was given the authority to revoke the winning banks' automatic right to sell the collateralized shares. In other words, the revised loans-for-shares scheme became an incentive for the banks to place their enormous financial, organizational and personnel resources at the service of Yel'tsin's re-election campaign to ensure a victorious outcome and consequently, their self-enrichment.[127] It was shrewdly modified by the 'young reformers' to serve as a crucial tool through which a coalition supportive of privatization reforms would be created and managed. In this way, the privatization of the oil industry was consolidated.

This interpretation of the loans-for-shares scheme, with an emphasis on its strategic dimension, is not without detractors who perceive it as an undisguised grab for assets.[128] Moreover, there is some doubt as to whether the strategic planning and foresight required of coalition-building was beyond the capabilities of Yel'tsin and his team who were too busy dealing with the economic issues at hand.[129] Nevertheless Yel'tsin and GKI had repeatedly demonstrated their ability for strategic thinking – Yel'tsin by appointing Gaydar's team as a means to maximize political power and GKI by making concessions on insider ownership to save the privatization programme as a whole. As such, it should not be surprising that loans-for-shares was used strategically as a coalition-management tool.

The loans-for-shares auctions were characterized by insider dealings and the use of legal and physical obstacles to deter outside bidders. As a result, the amount of credits that the state obtained during the trust auctions for oil companies (US$512 million) was at or just above the minimum bid for shares to be placed in trust management, even given that the minimum price had already been set below market rates. Furthermore, the winning bids at the collateral auctions in 1996 were not significantly higher than at the first auction, which meant that the state gained only an additional US$6.9 million, or 70% of the capital gain, from the sale of shares in oil companies.

In the case of LUKoil, proxy companies were used to secure the winning bids. The impact of an insider winning the collateral auction soon became clear. In June 1997, according to LUKoil's share quoted on the stock market, the 5% collateralized share was valued at US$608.3 million. The state therefore stood to gain US$401.31 million (or 70% of the capital gain).[130] Instead it received only US$6.01 million because the winning bid by LUKoil's affiliate was manipulated and lowered to only US$43.6 million.

As for Surgutneftegaz, 40.12% of the state's share in the VIOC was at stake and the auctions were won by its affiliates. In the trust auction held in November 1995, the bidding commission overseeing incoming bids comprised officials from Tyumen' *oblast'* (where Surgutneftegaz is located) and the general manager of Surgutneftegaz. Unsurprisingly, they disqualified the competing bid by state-owned oil company, Rosneft'.[131] To ensure that no further outside bids would be placed, the airport serving the remote auction site in western Siberia was closed for two days before the auction. Moreover, legal conditions set down by auction rules – which obliged the winner of the first auction to pay the company's tax arrears of US$223 million to the state, and required the winner of the second auction to transfer 5% of shares from the subsidiary to the holding company – worked in favour of Surgutneftegaz since outside bidders would be unwilling or unable to fulfill these requirements.[132] Finally, the fact that the two remaining bidders were affiliates of Surgutneftegaz left no doubt as to the outcome of the auction.

In the case of Sibneft', the winners of the initial auction, Neftyanaya Finansovaya Kompaniya (NFK) and Stolichny Bank Sbereshenii (SBS), were closely linked to the oil company. Two other bids by non-affiliated companies to manage the state's 51% share in Sibneft' were disqualified, and it was later revealed that they had each offered more than US$170 million, which far exceeded the winning bid of US$100.3 million.[133] During the second auction in May 1997, Potanin's ONEKSIMbank made a bid for the 51% of collateralized shares by placing the required deposit into the account of the organizers, SBS-Agro. Not unexpectedly, the organizers disqualified the bid on the basis that the required documentation accompanying the deposit had not been received. As such, an offshoot of NFK won the auction.

The auctions for YUKOS were arguably the most contested loans-for-shares transactions for oil companies, as the case study will explain. Following the victory of Khodorkovskiy's company in auctions held during 1995 and 1996, YUKOS became the first privately owned VIOC in Russia.

Apart from the loans-for-shares auctions, another method of privatizing oil companies was through a more standard method known as an investment tender. Whereas loans-for-shares auctions related to the cash sale of the state's block of shares (usually 45–51% of the VIOC) that was supposed to be held for three years, investment tenders were used to divest the state-owned portion of shares that were held by the investment fund of the VIOC, in exchange for investments on the part of the buyer to secure the long-term viability of the oil company. Although many of the investment tenders shared the characteristics of loans-for-shares auctions – insider deals at low prices, use of legal procedures to prevent unwelcome outside bids – on the whole they generated relatively less controversy and received less public attention. This was because the blocks of shares released for investment tenders were generally smaller in size than those at loans-for-shares, where control of the oil companies were at stake.

For example, Sibneft' held three separate investment competitions in January, September and October 1996 to sell 14%, 19% and 15% of its shares respectively. The proceeds from the sales amounted to over US$80 million and were to be used for a combination of write-off debts and equipment modernization.[134] The winners, in all cases, were companies affiliated to Sibneft' and its management who had won the trust auctions in 1995. Hence, by mid-1997, Sibneft' and its partner companies owned almost all of its own shares.

In the case of SIDANKO, legal requirements for an investment tender of 34% of shares in September 1996 and for the sale of collateralized state shares in January 1997 favoured ONEKSIMbank. In both instances, bidders were obliged to transfer a major stake in a subsidiary company, Angarsk oil refinery, to the holding company. However, only the Interros group, owned by ONEKSIMbank, possessed the necessary stakes to fulfill the conditions.[135] As such, ONEKSIMbank ended up with 85% of SIDANKO's shares.

Turning to TNK, the investment tender in July 1997 for a 40% stake in the company was won by Russia's Al'fa bank for US$820 million. The purchase was disputed by Viktor Paliy, president of TNK's subsidiary and largest oil production unit, Nizhnevartovskneftegaz, who alleged that Al'fa bank's victory had been predetermined as a result of collusion between officials from GKI, the federal bankruptcy agency (to which he owed back taxes), and some of TNK's board of directors.[136] In turn, Paliy was accused of failing to pay the subsidiary's taxes and of enriching himself by corruptly selling the subsidiary's oil abroad. He was sacked in September 1997, but continued to threaten Al'fa bank's attempt to consolidate TNK's shares. In the end, Al'fa bank won cash in the auction-cum-investment tender for the state's remaining stake of 49.8% for US$66.7 million in December 1999.

In short, by the end of 1997, the formerly state-owned oil industry was effectively in private hands. Indeed, among the top five VIOCs which together accounted for 55% of Russia's oil production in 1997,[137] LUKoil was the only one in which the state had any shares. The only major oil companies that remained under the control of the state were VIOCs such as Rosneft' and Slavneft', small production units such as ONAKO and KomiTEK and entities such as Tatneft' and Bashneft', which were partly owned by regional authorities.

The divestment of state property in the oil industry continued during Yel'tsin's second term and also under Putin. For GKI, the difference was that contrary to past practices when compromises were necessary to prevent the communists from winning the presidency, now it was necessary 'to tear the umbilical cord which binds together the state and business' so that 'the state would have no favourites and stepsons among commercial structures, but with everyone there would be equal, regular relations subordinated to clear rules'.[138] In the case of privatization, this meant that 'the first goal was to get as much money as possible for the state's assets'.[139] However, while the ensuing privatizations did fetch higher prices for the state, they were still not completely transparent or free of scandal.

This was because divestments of oil companies were still subjected to larger political considerations. In September 2000, an 85% stake in ONAKO was sold for a record US$1.08 billion, far in excess of the US$430 million starting price fixed by a Western investment bank organizing the sale. The winning bid was the largest sum at that time paid for an oil company that was not even a major player: in comparison, Menatep bank had obtained 78% of YUKOS for just over US$310 million between 1995 and 1996. As a result, the sale of ONAKO was hailed by state officials as 'honest, open and absolutely competitive'.[140] However, the competition was not as open as it was claimed since foreigners were excluded. Moreover, the fact that Al'fa bank outbid Sibneft' (controlled by Berezovskiy and his associates) and that this coincided with Putin's campaign against Berezovskiy, led to suspicions that the deal was rigged to benefit Putin-era oligarchic structures such as the Al'fa group, at the expense of holdovers from the Yel'tsin-era.[141] Al'fa bank had, in fact, opposed the loans-for-shares scheme proposed by the ONEKSIMbank-led consortium and was unsuccessful in its attempts to purchase stakes in YUKOS during the auctions.

In addition, the privatization of oil companies continued to be arranged in such a way that winning bids were usually predetermined. In October 1999, for instance, in an investment tender for a 9% stake in LUKoil, the winner was required to complete the construction of a semi-submersible drilling rig owned by Nikoil.[142] Given that Nikoil was an affiliate of LUKoil, an outside investor would have been hard pressed to fulfill this condition. Unsurprisingly, the only two firms that participated in the competition were linked to LUKoil. In any case, confirming the trend of higher prices for state assets, the 9% share was sold for over US$440 million: by way of comparison, the state's 5% stake in LUKoil had been sold for a mere $43.6 million at the loans-for-shares auctions in 1997.

Another way to predetermine winning bids was to appoint or co-opt management personnel in support of the interested buyer. In this way, the profits and revenue flows of a state-owned company could be privatized, which would ensure the buyer a strong position as a leading bidder.[143] In the case of Slavneft', for example, Sibneft' nominated a number of its top executives to join the management team at Slavneft', while its oil trading subsidiary was appointed to handle the exports of Slavneft'. Thus in December 2002, Sibneft' and its ally, TNK (part of the Al'fa group of companies), successfully purchased all of the state's 75% share in Slavneft'.[144]

Policy consolidation in Russia

The successful political management of privatization by pro-reform groups in the government was the crucial reason why privatization in the oil industry was consolidated, at least prior to 2003.[145] One aspect of political management of privatization involved the creation and maintenance of a coalition of allies supportive of privatization and of the incumbent administration. In this connection, the 'young reformers' initially accepted insider ownership by management and employees of corporatized oil companies on the assumption that any private owner was better than the state. Later, they skillfully modified Potanin's loans-for-shares proposal to exchange the state's block of shares in lucrative companies in return for political and financial support from influential corporate leaders. In addition, as will be explained later, the 'young reformers' also took measures to ensure that the Duma did not become a significant obstacle to the consolidation of privatization in the oil industry.

It was previously argued that Yel'tsin had delegated some of his decision-making prerogative to a team of technocrats because he calculated that they would institute the kind of anti-Soviet, pro-market reforms that were significant for his electoral constituency and hence, his political longevity. The opposite is also true: that the reason why pro-reform groups during the Yel'tsin administration were able to successfully manage the political aspects of privatization, despite having almost no power base of their own, was largely due to support from the executive. It is true that Yel'tsin was not always publicly supportive of the liberal reformers; although he enlisted their help in 1992 and turned to them again between 1997 and 1998, he replaced them with conservative governments in the intervening periods.

Nevertheless alternating between periods of 'breakthrough' and stabilization helped Yel'tsin to consolidate his own position and consequently, the spread of privatization in the oil industry.[146] This is because the purposeful shifts reinforced his role as an indispensable leader who occupied both the centre and radical wings of the Russian political spectrum. At the same time, this pattern of change followed by stasis allowed the gains from privatization to be consolidated more easily than a strategy of constant change, which has historically resulted in violent upheavals and policy reversals in Russia. Yel'tsin himself admitted that there was a certain logic to what appeared to be idiosyncratic changes. Accordingly, 'in order to preserve the status quo, I had to bring in new figures from time to time. I had to promote some people and sacrifice others. But any sacrifice, any dismissal, any change of the political configuration could not be accidental or merely tactical. In each of my moves, I had to keep the overall strategy and main purpose in mind'.[147]

In Putin's case, it suffices to mention at this point that his support for divesting state-owned oil companies to the private sector has depended entirely on his calculation of how they affect his political survival and authority in Russia. This is because, as Yel'tsin before him, Putin is a pragmatist who is not swayed by political ideologies. Yel'tsin was 'prepared to tolerate democracy as long as it did not threaten his position and he was prepared to renounce or ignore the most important principles of democracy ... in defense of his own power'.[148] As for Putin, he is strongly in favour of the need for Russia to 'search for its own path to renewal' and 'not become a second edition of, say, the US or Britain' or indulge in social experiments such as communism.[149] He has been described as a 'liberal *chekist* ... his economic vision is largely pragmatic and liberal, relatively open to the market and to foreign investment – if only the best tools to rebuild Russia's influence in the world'.[150] The fact that Chubays has remained the head of Unified Energy Systems is a testament to Putin's belief in 'what works', as the following remark by Putin illustrates: 'I thought that [Chubays] lived in a world of illusions. But it turned out that he's more of a pragmatist, that he's capable of grasping the realities of life and is not guided by ephemeral ideas.'[151]

Given Putin's pragmatic nature, if it was his assessment that a particular case of divestment enhanced his political longevity, then supported of privatization, as was the case with ONAKO and Slavneft'. In both these cases, the process was managed to ensure that the companies were purchased by business groups that did not threaten Putin's still fragile hold on the presidency at that time, that is, those persons who were 'more businessmen than politicians';[152] they were modest in lifestyle, loyal and refrained from engaging in overt political activities.[153] If, however, his political future was enhanced by increasing state-ownership in the oil sector, then he will supported de-privatization, as exemplified by the YUKOS affair. As will be explained in the case studies, Putin's behaviour with respect to YUKOS and Rosneft' was predicated on his own coalition management strategy, in much the same way in which loans-for-shares was used by Yel'tsin and his 'young reformers'.

Another way in which the shrewd political management of privatization shaped the contours of privatization in the oil industry concerned the intensity of implementing change, particularly, in view of the approach of parliamentary and presidential elections. The expectation that the Communists would perform well at the December 1995 parliamentary elections motivated the 'young reformers' to accelerate the divestment of state-owned shares. Indeed, 'if the only way to get Russia's leading energy and transport companies into private hands before the December 17, 1995 elections was through the loans-for-shares programme . . . so be it'.[154]

Subsequently, the 'young reformers' concluded that post-electoral developments were symptomatic of the 'changed alignment of political forces'[155] in the country. The Communist Party and its ally, the Agrarian Party, for instance, increased their share of the popular vote from 20.3% to 30.6% between 1993 and 1995. On the other hand, liberal parties recorded a fall in support from 34.1% of the popular vote in 1993 to 16.1% in 1995.[156] Worse for Yel'tsin, his approval ratings slumped to just 8% in January 1996 against 21% for Zyuganov, and although they subsequently rose, Yel'tsin still trailed Zyuganov.[157] These results were interpreted as a reversal for reform[158] and by way of a tactical retreat, Chubays stepped down as GKI Chairman in favour of Aleksandr Kazakov. The latter promptly announced an end to collateral auctions, the introduction of a slower and more individualized phase of privatization and his intention to participate in the Duma's Commission to Analyse the Results of Privatization.[159] At the same time, investment competitions to sell-off significant blocks of shares in oil companies were put on hold, and it was not a coincidence that they only resumed in September 1996, after Yel'tsin's victory in the presidential elections. Hence, although the parliamentary elections did not deter – but in fact encouraged – the 'young reformers' from proceeding with politically unpopular loans-for-shares transactions, the perceived impact of those elections on Yel'tsin's campaign compelled a relaxation of privatization, which is consistent with existing case studies on the impact of elections on reform consolidation.

Putin was also equally shrewd with regard to elections and implementing changes in privatization. Many analysts are of the opinion that the YUKOS affair was a direct consequence of Khodorkovskiy's political ambition and the challenge this posed to the political status quo.[160] Khodorkovskiy himself encouraged assumptions about his political aspirations. For example, he remarked that 'at 45, I will no longer want to be an economic leader and then will perhaps go into politics' and noted that he 'never said that big business must keep itself outside of politics'.[161] Khodorkovskiy was also financing opposition parties and candidates. For instance, up to half of Yabloko's annual budget was underwritten by Khodorkovskiy, who had donated to the party since 1993.[162] Businessmen with ties to YUKOS were also represented on the party lists for Yabloko and the Communists. Nevertheless other tycoons were also supporting opposition candidates, albeit to a lesser degree.[163] Yabloko was, in fact, never a credible threat to the dominance of the pro-Kremlin United Russia, as it was only expected to win 6–7% of the vote in the December 2003 elections.[164] In this regard, the prospect that the liberal parties would together dominate the Duma in 2004 and force through a constitutional amendment transforming Russia into a parliamentary republic by 2008, with Khodorkovskiy as prime minister, was unrealistic.[165] Therefore, while Khodorkovskiy's electoral activities were irritating for Putin, they were unlikely to have been a source of major concern for the president.[166]

Instead, Putin used the YUKOS affair as an election campaign tactic to drum up maximal support for United Russia and discredit opposition parties. In this connection, one Russian newspaper opined that 'every presidential campaign in Russia has required a public enemy, and the upcoming campaign is no exception. In 1996 we fought against the communists, in 1999 we fought against the Chechen terrorists. In the presidential campaign of 2004, it is clear that we are going to do battle against the oligarchs'.[167] By making a campaign issue of the questionable origins of the wealth of leading businessmen, Putin opened a policy window to review past privatizations, in particular, of oil companies. As will be explained later, this allowed Putin to consolidate his power vis-à-vis big business and to embark on his plan to create energy behemoths controlled by his allies. In addition, with United Russia in control of more than two-thirds of the seats in the Duma, Putin was able to use the Duma to rubber-stamp his policy initiatives.

Policy consolidation in the Russian oil industry also benefited from the weakness of opposition groups. The non-oil industrial, military and agricultural sectors were relatively unsuccessful in affecting the progress of oil privatization for several reasons. Particularly during the first half of the 1990s, the state was unable to financially support the demands of the oil consuming sectors for cheap oil because it was focused on restoring monetary stability to the economy. Furthermore, attempts by GKI to co-opt organized interests into supporting privatization were successful. For instance, general managers of leading oil companies who were initially opposed to privatization later grabbed the opportunity to own their enterprises either through their own affiliated companies (in the case of LUKoil, Surgutneftegaz and Sibneft') or in concert with banks (in the case of YUKOS and TNK). Similarly, the incentive for defence-related companies to continue to support the anti-privatization group was reduced by a presidential decree issued in December 1995, which banned the privatization of enterprises within the military–industrial complex.[168]

As for the parliament, its role in the policymaking process with respect to privatizing the oil industry cannot be reduced to one of powerlessness or 'fig-leaf parliamentarism'.[169] Nevertheless it has not been one of the key actors in the process. As noted earlier, the major privatization-related legislation, as well as the creation of VIOCs, took the form of presidential decrees and government resolutions. Under Yel'tsin, the parliament's influence was largely limited to delaying the pace of privatization, but even in this respect, its successes were largely contingent on the fact that decisions to postpone privatization had already been taken by more influential actors. Following the December 1995 parliamentary elections, for example, the Communist Party managed to obtain the resignation of Chubays and a more conciliatory policy on privatization from his successor at GKI, Kazakov. However, as previously mentioned, the Duma did not trigger this response from GKI, since such a tactical retreat ahead of Yel'tsin's re-election in mid-1996 had already been planned by the latter. The Duma's most consequential intervention in oil privatization was arguably its ban, enacted in mid-1995, on the sale of the state's shares in companies producing strategically important goods. This prohibition certainly made the loans-for-shares an attractive proposition, since credits could be obtained for the budget without having to actually divest of the state's shares in such companies. Nevertheless, as argued above, loans-for-shares was not the result of the Duma's said ban but, rather, of a coalition management strategy in the interests of Yel'tsin and his 'young reformers'.

Under Putin, the Duma managed to pass legislation that prevented the sale of Slavneft' in 2001. However, as the case study will explain, the initiative for this move originated not from the Duma itself, but from the president of Slavneft' who used his former position as a deputy speaker of the Duma to postpone privatization for his own benefit.[170] Likewise, although the request for the Prosecutor-General to investigate the privatization practices of YUKOS originated from the Duma in 2003, its degree of success was dependent on the political agenda of more influential actors in the presidential administration. In comparison, the Duma's request for the Prosecutor-General to investigate the privatization of TNK and Sibneft' has gone unanswered.

One reason for the parliament's relative lack of influence in privatization concerns its institutional legacy. The Soviet-era parliament, which had allied itself with those organized interests opposed to privatization, found its powers severely curtailed as a result of the December 1993 referendum. The superpresidential regime which was established thereafter used its powers of decree to initiate and perpetuate privatization in the oil industry.

Another reason relates to the fact that Duma deputies have been susceptible to bribery and other forms of inducements from tycoons keen to acquire oil companies and from GKI itself. According to a parliamentary deputy, the 'mechanism of lobbying in the Duma is primitive: you simply buy votes'.[171] Sometimes, inducements were more subtle, as was the case in Samara region: having successfully managed the incumbent governor's re-election campaign in 2000, YUKOS's vice-president was duly appointed deputy governor while the governor's son was given a position in a YUKOS-affiliated bank.[172] In the case of GKI, one of its former deputy ministers responsible for managing relations with parliamentary deputies revealed that deputies opposed to privatization 'always had some property interests – like they needed an office for their party, or there were local property interests to protect in their own constituencies. So it was always possible to come to a consensus with some of them'.[173]

Moreover, the lack of coordination between the legislative agenda of the executive and parliament resulted in conflicting policy priorities, especially during Yel'tsin's presidency. This was the consequence of an 'absence of a stable, party-based majority in the Duma, the fact that governments are not party-based, and that neither [Yel'tsin] nor Putin were elected on party platforms'.[174] This means that the executive has tended to bypass the parliament when it is unable to persuade the parliament to approve its policy agenda.

As for the banking interests that desired to gain control of oil companies, they clearly played an active role in consolidating the progress of privatization through proposing the loans-for-shares scheme, negotiating with some of the 'oil generals' to form a strategic partnership, buying support from regional elites and colluding with GKI officials to discourage unwanted bidders. Moreover, business corporations and oil companies supportive of further privatization also ensured that their interests were represented in parliament. For example, an interparty group known as Russia's Energy was created in February 2000 with Chernomyrdin as its chairman, and claimed that one of every five members of the Duma owed their allegiance to the oil and gas sector.[175] In this regard, it is hardly surprising that the Duma has, for the most part, not been a more vociferous opponent of privatization. The interests of oil companies have also been represented in the upper chamber of parliament, the Federation Council, where many oil executives were appointed as senators by regional elites keen to benefit from the largesse of oil companies. For instance, Roman Abramovich, who controls 80% of the shares in Sibneft,[176] is the governor of Chukotka, and his former colleague in Sibneft', Yefim Malkin was elected as a senator for the region. Leonid Nevzlin, co-owner of YUKOS and co-founder of Menatep bank, was a senator for Mordovia, where YUKOS subsidiaries are based, while his colleague Boris Zolotarev, was governor of Evenkiya.[177] In addition, as mentioned earlier, the leaders of large business corporations took control of the leading business organization, the Russian Union of Industrialists and Entrepreneurs; as a result, the organization was transformed from one that opposed radical reform to one that was supportive of pro-business policies, including privatization.

Conclusion

This chapter has argued that the initiation and spread of privatization in the Russian oil industry after 1992 was primarily a result of conscious and repeated choices by key political and economic actors within Russia. Specifically, the initiation of decree 1403 and the creation of VIOCs were driven by leading *nomenklatura* managers of oil entities and key state officials from GKI and Mintopenergo. They behaved partly out of an attempt to maximize their respective self-interests and also in response to their perception of the legacies of the Soviet past. In this respect, the rather simplistic notion that privatization was the result of capture by tycoons or 'oligarchs' should be reconsidered, since they were clearly not actors of consequence in the oil industry prior to 1995. The material self-interest of the tycoons was a key variable in the spread of privatization after 1995, but even then, they were not the only ones driving policy; the *nomenklatura* general managers and key state officials continued to be deeply involved in managing and consolidating privatization in the oil industry. Since the flow of benefits accrued to all three actors concerned, that is, their respective self-interests were maximized or at least satisfied, the basis of their interaction was arguably one of mutual exchange rather than an outright capture of benefits by any single participant.

Other factors that are generally assumed to account for policy initiation and consolidation – including crisis situations, expert communities, the internationalization of privatization – formed the parameters around which decisions were made, but did not necessarily predetermine or trigger the actual choices adopted by the key decision-makers.

Having analysed the policy process on a macro or industry-wide level, the subsequent chapters will allow a detailed, company-level analysis of the privatization experiences of YUKOS, Slavneft' and Rosneft'. The focus will be on how, why and to what extent the pursuit of rational self-interest was reinforced, enhanced or even constrained by institutional, historical and international circumstances.

3
Case Study of YUKOS

Accounts of the rise and fall of YUKOS typically begin with the collateral auction of December 1995, and the subsequent sale of these shares together with an investment tender a year later, which gave Khodorkovskiy's Menatep group control of over 85% of YUKOS's shares.[1] However, this focus on Khodorkovskiy mistakenly ignores the significance of the preceding years when YUKOS was founded and corporatized. During this time, the key roles were played by former members of the *nomenklatura*, whose ability to shape the development of YUKOS was a result of institutional legacies and incentives, as well as of personal choices and preferences. In contrast, Khodorkovskiy played a key role in the actual privatization of YUKOS in 1995 and was also central to the YUKOS affair of 2003–2004.

Corporatization by the *nomenklatura*

At the end of 1991, discussions on the creation of YUKOS were initiated by Sergey Muravlenko, the general manager of Yuganskneftegaz, a major oil producer, and by Vladimir Zenkin, the general manager of Kuybyshev oil refinery. Their decision to merge the two state-owned entities into a single holding company, YUKOS, was influenced by several factors. First, as noted in the previous chapter, the disarray in the Soviet oil industry encouraged the general managers of state-owned enterprises to rapidly expropriate state assets for themselves before the window of opportunity closed for good. Since 1991, for instance, Zenkin had been exporting refined oil products through a joint-venture with a Belgian company, and had made enormous profits due to the huge discrepancy between domestic and world market prices for such products.[2] The last Soviet Oil Minister Lev Churilov recalled that he had received an endless series of requests and directives from senior government and Communist Party officials to allocate particular oil fields to specific companies with which they had links.[3] Such behaviour by the *nomenklatura* in fact accelerated the collapse of Soviet institutions, since as a 'shrewd and mercenary gravedigger ... it aimed to profit from its own death'.[4] Acquiring legal control would therefore enhance the *de facto* property rights of *nomenklatura* managers over their respective production and refinery enterprises. Muravlenko and Zenkin, as president and vice-president, respectively, would then be in a position to run the merged company, YUKOS, with minimal interference from the state and, in the process, further enrich themselves.

Secondly, Muravlenko and Zenkin were inspired by Alekperov. As noted in Chapter 2, Alekperov gathered together three hitherto separate oil producers to form the LUKoil state oil *kontsern*. With the dissolution of the USSR, he began negotiating with other downstream entities (that is, those engaged in refining, marketing and distribution) to create a VIOC which was the norm in other oil-producing countries but a novelty in Russia. Seeing what he was doing, the general managers of other oil entities 'just followed him [Alekperov] and tried not to miss the opportunity'[5] to create and control their own oil enterprises. Therefore, when Muravlenko and Zenkin learned that Alekperov was lobbying for a decree to officially recognize the creation of LUKoil as a joint-stock company under Russian law, they supported the idea and 'using their connections with top economic and political authorities in the country, they were able to get their companies included in decree 1403 at the very last moment'.[6] Muravlenko himself acknowledged the significance of Alekperov's role when he remarked that because LUKoil was the first oil *kontsern*, it had already 'amassed some experience. We [YUKOS] follow in their trail and I must hand it to Mr Alekperov that he always shares his expertise and knowledge with us'.[7]

The two factors discussed above formed the general institutional parameters within which the oil industry operated at the beginning of the 1990s. What specifically brought Muravlenko and Zenkin together was their personal relations and the perceived synergy between their enterprises. Muravlenko was, in fact, born in Kuybyshev, the same town where Zenkin's Kuybyshev refinery was located. Moreover, Muravlenko's father, Viktor, had worked in the oil industry there before going on to become a legendary leader within the Soviet oil industry.[8] Muravlenko was therefore quite closely connected with the Kuybyshev region and was able to trade on his family's name, reputation and contacts.

Muravlenko and Zenkin were also united by the synergies of their companies. Yuganskneftegaz possessed the largest amounts of oil reserves and was the second-largest oil producer in the Soviet Union, accounting for 11% of Russia's total output in 1991.[9] As for Kuybyshev, it was a leading refinery with the capacity to produce a wide range of petroleum products and, more importantly, its refinery capacity was large enough to absorb most of the output of Yuganskneftegaz, the balance of which could be exported. The ability to keep production and refining operations within the same company meant that YUKOS would be able to reduce costs by not relying on third-party refineries.[10] Yuganskneftegaz had, in fact, rejected outright an earlier proposal to merge with Yaroslav refinery as the latter's capacity was insufficient.[11]

The creation of YUKOS was thus an exception to the general rule that 'decisions regarding which subsidiary enterprises were to form which VIOCs were based less on commercial or economic rationality than on personal connections and power plays'.[12] Indeed, according to senior oil industry executives, the original joint-stock companies such as YUKOS, 'were selected based purely on technological principles',[13] while in many of the later oil companies, 'the balance between production, refining and sales . . . became distorted'.[14] For example, the refineries in Rosneft' and SIDANKO were located far from the main production fields thus increasing transportation costs, while ONAKO and Tatneft' were basically production entities disguised as integrated companies since they had almost no refining capacity. Hence, while existing institutional incentives within the oil industry and personal connections certainly played a role in the creation of YUKOS, the objective, technological complementarity between Yuganskneftegaz and Kuybyshev was just as significant.

Muravlenko and Zenkin thus joined Alekperov in lobbying for official recognition of vertically integrated companies in the oil industry. At the same time, the two of them invited other distribution, marketing, transport and refinery enterprises to join in the discussions, the outcome of which was the formation of YUKOS as an open joint-stock company (or corporatized entity), which derived its name from its two major constituent parts, YUganskneftegaz and KuybyshevnefteOrgSintez. This was approved on 11 November 1992 by presidential decree 1403, which also endorsed the creation of LUKoil and outlined a new structure for the Russian oil industry in which corporatized and vertically integrated companies would dominate. The subsequent Government Resolution 354, issued on 15 April 1993, marked the start of operations at YUKOS and also approved the company's privatization programme. The change in status for YUKOS provided legal recognition of control by Muravlenko and Zenkin over the company, since only a 45% stake was retained by the government. Any potential threat to their position posed by foreign owners had also been minimized thanks to a provision contained in this same decree that limited foreign share ownership in a holding company, such as YUKOS, to 15%.

Muravlenko and Zenkin therefore made good use of the institutional legacies of the Soviet period – their managerial position, network of political connections, *de facto* ownership rights, and economic and legal disarray of the early 1990s – to become members of the post-Soviet business elite. Indeed, the fact that 61% of the Russian business elite in 1995 was drawn from the ranks of the Soviet *nomenklatura*, appeared to lend support to those who argued that economic reforms in Russia had resulted in a perverse form of capitalism known as *'nomenklatura capitalism'*.[15]

At the end of October 1993, the management of YUKOS implemented the privatization plan for its main production subsidiary, Yuganskneftegaz. 12% of Yuganskneftegaz's shares were offered to the public in exchange for privatization vouchers with a face value of 10,000 rubles that had been allocated to every Russian citizen as part of the mass privatization scheme launched the previous year. The rest of Yuganskneftegaz's shares were divided among its workers and managers (44%), its parent company, YUKOS (38%), and oil transportation employees and indigenous people living in western Siberia (6%), where the enterprise was located. Hence, by the end of 1993, almost 15% of the share capital of YUKOS had been privatized.

From the outset, YUKOS faced two key challenges. The first was to raise capital for investment and modernization in order to reverse the decline in oil production levels. Between 1992 and 1995, YUKOS's share of production in Russia held steady at around 11% of total output. Nevertheless the company's annual output declined by nearly 50%, from 53 to 36 million tonnes during that period, which, in fact, reflected industry-wide trends. The World Bank estimated that US$50–US$60 billion would be required for the oil industry merely to stabilize production levels between 1995 and 2005. The problem was that total foreign direct investment in Russia as a whole amounted to a paltry US$5.5 billion by 1995,[16] and the state was in no position to underwrite the financial needs of the oil industry. Even a loan from the World Bank to modernize oil production facilities failed to make up for the shortfall.[17]

The other key challenge was to fully integrate YUKOS's subsidiaries, some of which were resisting efforts to merge their operations into the larger holding company, because this would reduce the opportunity for asset-stripping at the subsidiary level. Viktor Ivanenko, a former vice-president of YUKOS (1993–1995), admitted that 'each leader at every level of management had their own business schemes which were basically considered criminal. This was a problem for all of YUKOS's subsidiaries and this happened not only in YUKOS but at other oil companies as well.'[18] Yuganskneftegaz, for instance, had established its own bank, Tokobank, to handle its oil exports, thereby bypassing the holding company's preferred banking partner, Promradtekhbank. In another YUKOS subsidiary, NovoKuybyshev refinery, the general manager, Viktor Tarkhov, brought a law suit in 1993 against the Russian government with respect to the latter's decision to include it within YUKOS earlier in the year. Strikes and walkouts supported by Tarkhov disrupted operations at the refinery. In reality, Tarkhov was unwilling to hand over the refinery's financial flows to the holding company because this would mean having to give up the money he was receiving from his foreign partners for illegal oil exports.[19] Worse for Muravlenko, his partner and YUKOS's vice-president, Zenkin, was murdered in late 1993 by the latter's former Belgian associates who perceived his attempts to place financial flows from Kuybyshev refinery under the control of YUKOS as a threat to their hitherto lucrative relationship.[20]

Muravlenko and the management of YUKOS failed to deal decisively with these twin problems not because they were powerless or ill-equipped to do so. As mentioned above, the general manager was, in fact, a very influential figure, thanks largely to the personalistic power inherited from the Soviet system. Indeed, 'Russian oil companies at that time had a very hierarchical structure – they were run only by one man',[21] usually the founder of the company. Rather, Muravlenko failed to respond adequately to these challenges because he was unwilling to do so. The state bureaucratic agencies that were given oversight of the oil industry, in particular Mintopenergo and GKI, were too busy developing and refining the structure of the industry between 1993 and 1995. Issues of corporate governance were dealt with only sporadically whenever particular weaknesses of the legal regime became apparent, and the first comprehensive corporate law was adopted only in 1996.[22]

In terms of work practices, Muravlenko had a non-confrontational style of management and chose to 'put aside the internal structural reorganization until an official decision of this issue',[23] unlike Alekperov who had already proactively and successfully integrated LUKoil's subsidiaries into the holding company. Moreover, Muravlenko was a paternalistic leader who, 'being a representative of the old Soviet school, still wanted to be kind. He didn't want to get rid of the social burden, especially in Nefteyugansk [the town where Yuganskneftegaz is located]. He had built a lot of hospitals there, an institute of technology and helped local schools a lot. Any wishes from Nefteyugansk's mayor's office were immediately fulfilled'.[24]

In addition, he was himself engaging in asset-stripping and had little incentive to impose financial discipline and transparency within YUKOS. Despite the fact that he owned 5–10% of the holding company's shares, his effective control over YUKOS enabled him to pocket the entire profit of the firm, while suffering only minimal losses from his own shareholding as result of his asset-stripping.[25] At that time, almost nobody believed that the state would actually privatize oil companies; from Muravlenko's point of view, there was hence nothing to be gained from attempting to enhance the value of YUKOS's or his own shares.[26] Institutional factors and personal motives therefore accounted for Muravlenko's reticence in resolving the problems with debts and subsidiaries prior to 1995.

In early 1995, however, Muravlenko began to approach banks with a view to a negotiated takeover. This was, in part, a response to increasing pressure from the government, which was alarmed by the high operating costs at YUKOS (partly due to the failure to rein in wayward subsidiaries) and the company's increasing wage, pension and tax arrears;[27] indeed, as Ivanenko, its vice-president admitted, YUKOS was 'almost bankrupt'.[28] The government's new activism was underlined when it transferred ownership of oil producer, Purneftegaz, from tax-delinquent SIDANKO to state-owned Rosneft' in 1994; it was rumoured that some of YUKOS's refineries, which had initially objected to being part of the holding company, could face the same fate.[29] YUKOS's viability as a company and its leadership capability therefore came under heavy scrutiny from the government. In order to protect his own position and source of wealth, Muravlenko made belated personnel changes at YUKOS's subsidiaries, but to little avail.

However, institutional pressure by itself is not a sufficient explanation for why Muravlenko chose to approach third parties for help with implementing far-reaching organizational changes. This is because as an alternative, Muravlenko could have effected a management buyout in response to pressure from the government to improve corporate governance and performance at YUKOS. This was the strategy adopted by LUKoil and Surgutneftegaz, where the core of the original management team currently controls 40% and 68% of shares in their respective companies.[30] However, it was not a realistic option for Muravlenko because 'he hadn't the ability or willingness or political power to privatise it by himself'.[31] He therefore would not have been able to provide the strong, decisive and confident leadership displayed by Alekperov and Vladimir Bogdanov of Surgutneftegaz that was crucial for a successful management buyout. As previously noted, Muravlenko had obtained his position in the Soviet oil industry mainly as a result of his father's contributions, and though well liked, he was not a brilliant manager.[32]

To Muravlenko's credit, he recognized his own limitations and regarded a strong partner as the only realistic way of continuing to profit from his shareholdings in YUKOS. Ivanenko himself acknowledged that 'these young specialists of Khodorkovskiy were a hope. I hoped that these people would be able to do something radical . . . as I saw that the old team was unable to do anything'.[33] Hence, while institutional pressures forced Muravlenko to come up with a strategy to improve performance at YUKOS, his decision to work towards privatization through an alliance with a suitable financial institution was the result of individual choice.

Muravlenko's behaviour is reminiscent of Mancur Olson's self-interested 'stationary bandit'.[34] Unlike a 'roving bandit' who pillaged and moved on to other targets, a 'stationary bandit' would leave some assets to generate future growth so that the enterprise could be pillaged over a period of time, thus increasing his total takings over and above what he would receive as a 'roving bandit'. Hence, while engaging in asset-stripping, Muravlenko simultaneously searched for a financial institution that was able to enhance the value of YUKOS and also that of his own shareholding. Why then did Muravlenko choose to be a 'stationary bandit' as opposed to a one-off asset-stripper, considering that the former required a relatively longer time horizon and thus entailed a higher risk of getting caught and of not being to liquidate or transfer assets abroad as quickly?

Two factors in particular appeared to have encouraged a longer-term perspective than that adopted by many other enterprise managers, the first being that he was motivated by some sense of duty to his familial reputation and legacy. Thus, Muravlenko felt obliged to attempt to generate some semblance of growth in YUKOS. The second key factor that encouraged a longer-term perspective was his growing contact with potential partners and the increasing likelihood of privatization in Russia, which meant that the value of the company could be traded and was no longer 'locked-in' as before. These new, informal practices provided a powerful incentive for him to minimize asset-stripping, which would negatively affect the company's valuation and the future value of his own shareholdings.[35] To underline its commitment to reducing the role of the state, a programme of share divestiture through cash privati-zation was announced in July 1994, the most well-known outcome of which was the loans-for-shares scheme.[36]

One of the earliest plans to consolidate YUKOS's control over its sub-sidiaries was drawn up by its banking partner Promradtekhbank in 1994. However, the bank's role was marginalized: it was merely a medium-sized bank and could not compete with the dynamism and aggression of the new, Moscow-based commercial banks.[37] Around the same time, YUKOS was approached by renowned foreign investment banks, such as MC Securities and Lazard, but their proposals were rejected as they did not adequately take into account the interests of YUKOS's senior management.

In early 1995, Muravlenko entered into discussions with leading Russian banks. Potanin's ONEKSIMbank was quickly given the cold shoulder by Muravlenko who felt slighted because Potanin did not show him enough respect: at one meeting, Potanin left his deputies in charge while he himself went hunting.[38] A three-bank consortium backed by Lazard, Paribas and Credit Suisse First Boston (CSFB) emerged as the key challenger to Menatep for YUKOS for several reasons. First, the consortium included representatives of the largest and most dynamic banks in Russia.[39] Second, the leaders of these banks – Vladimir Vinogradov (Inkombank), Mikhail Fridman (Al'fa bank) and Vitaliy Malkin (Rossiyskiy Kredit) – had been excluded from the group of tycoons that had proposed the loans-for-shares scheme, and they were thus eager to assert themselves. Third, its Western backers had close ties to the Russian government: the head of Lazard's Moscow office was Lopukhin, while CSFB had worked closely with GKI in designing the mass privatization scheme.

Faced with several suitors, Muravlenko decided to ally himself with Menatep for several reasons. First, Khodorkovskiy offered Muravlenko and his closest associates at YUKOS a deal they could not refuse in terms of money, share options and employment. Secondly, Muravlenko was 'attracted by the lobbying capabilities of Menatep',[40] which had the sup-port of key figures in the state bureaucracy for Menatep to acquire YUKOS. They included Deputy Prime Minister Oleg Soskovets and Deputy Finance Minister Aleksandr Samusev. Third, the fact that the Menatep-led consortium comprised only native, Russian banks and companies – all of which had business dealings with YUKOS – appealed greatly to Muravlenko's sense of civic pride in Russia and in the oil industry.[41] Alluding to his patriotism, a former colleague from YUKOS suggested that Muravlenko decided 'to choose first a Russian financial group, and then one with big, lobbying capacities'.[42] Indeed, at the first shareholders' meeting after winning the investment tender held in December 1995, Menatep, its bidding partners and YUKOS jointly declared that they would not sell, pledge as collateral or transfer the shares to a foreign company.[43] This sense of civic pride was again demonstrated when YUKOS explained that in November 1995, it had rejected a bid by Inkombank's consortium because 'a large part of the money would come from foreign sources ... this means that a controlling block of shares would go abroad'.[44]

Privatization of YUKOS by the new elite

Thus far, the chapter has analysed the parameters of the decision-making process at YUKOS, particularly the institutional, material and other (familial and civic pride, as well as his own limits as a successful manager) influences on Muravlenko. He clearly played a pivotal role in YUKOS's formation, corporatization and search for an external shareholder, a fact which is often overlooked. We will examine next Khodorkovskiy's role, motivation and method of privatizing YUKOS. The words 'Khodorkovskiy' and 'Menatep' will be used interchangeably since, at that time, he was the group's founding president and owned 60% of Menatep's shares at one point.[45]

Khodorkovskiy was interested in acquiring assets in the oil sector primarily because of the huge rents that the industry offered as a result of the significant price differential between the domestic and international price of oil. As explained in the previous chapter, his interest in the industry grew from late 1994 onwards when it was becoming apparent that the conditions under which Menatep bank had in the past greatly profited currency speculation, trading in government bonds and 'authorised' banking status – were changing. Khodorkovskiy was particularly keen to acquire YUKOS partly because it was the second-largest company in the country after LUKoil in terms of output and reserves. In addition, YUKOS was 'available' – that is, it was more vulnerable to an external takeover than other leading oil companies in view of weaknesses in its management, which made no secret of the fact that it was keen to establish a strategic alliance with an external partner. LUKoil and Surgutneftegaz, for example, had strong management teams, Sibneft' had already been singled out for acquisition by Berezovskiy, while deeply indebted SIDANKO was engaged in advanced negotiations with its creditor, ONEKSIMbank.

As indicated above, Khodorkovskiy's role in the privatization of YUKOS began prior to the introduction of the loans-for-shares scheme.

His negotiations with Muravlenko to create a strategic alliance between Menatep and YUKOS were in anticipation of the end of the three-year period (that is, in April 1996), when the state was due to divest its share-holdings in YUKOS in accordance with the privatization programme for the company adopted at its creation. Despite the fact that Yel'tsin subsequently issued a decree in May 1995 ordering the government to explore the loans-for-shares scheme, Khodorkovskiy was not convinced that the scheme would actually be implemented, and as such, he actively initiated an alternative strategy aimed at achieving his goal of acquiring YUKOS. In this connection, he proposed an equity swap, whereby the government was offered a 10–16% stake in Menatep bank in exchange for the state's 38% shareholding in YUKOS.[46] However, this proposal was rejected in view of the government's acute budgetary needs – by June, GKI had only raised 104 billion rubles from cash privatization against a target of 8.7 trillion rubles for 1995 – and the fact that the swap would not result in a cash injection into privatization receipts.

The loans-for-shares scheme, which was finally approved in August 1995, therefore presented Menatep with the most likely strategy to realize its alliance with YUKOS. Khodorkovskiy and his associates actively manipulated the legal conditions of the trust auctions of December 1995 in such a way as to dissuade rival bidders from participating. As such, rather than engaging in 'state capture' which involved simply corrupting state officials to draw up preferential rules for the auction, Menatep itself directly participated in creating the new rules of the game. In so doing, Menatep reiterated previous empirical studies which have shown that large firms, whether private or state-owned, consistently pay a smaller percentage of their total revenues in bribes as a result of their pre-existing, overwhelming structural power in the economy and network of political connections. These factors render large firms more efficient bribers; they can purchase more influence per dollar of bribe compared to smaller and less economically significant companies.[47]

First, Menatep managed to get itself appointed as the auction organizer for YUKOS, which gave it the right to set the conditions governing the auctions.[48] Accordingly, Menatep required each bidder to deposit US$350 million into the Central Bank. It explained that the size of the bond was due to parallel auctions for the company's stock on the same day: a 'trust auction' for management of the state's 45% share in YUKOS, in addition to an investment tender for 33% of the company's stock. The size of the deposit was, in fact, intended as a deterrent to potential bidders since it exceeded the statutory capital of most Russian banks. For example, ONEKSIMbank, one of the largest in the country, had an equity capital of only US$550 million while equivalent figures for Inkombank, Al'fa bank or Menatep were far smaller.[49] Questions were also raised about Menatep's sources of funds which were alleged to have been accumulated by YUKOS as a result of its tardiness in paying federal taxes. Others speculated that Menatep may have used funds which its largest client, the Ministry of Finance, deposited into the bank, so that the amount of money that actually changed hands during the YUKOS auction was either minimal or non-existent.[50]

As auction organizer, Menatep imposed a second condition to further limit the circle of potential bidders – foreigners were excluded from participating in YUKOS's auction. Although the decree of August 1995 did not specifically exclude foreigners from participating in the loans-for-shares auctions, Menatep used its discretion as auction organizer to impose this ban as an additional condition.[51] This condition was personally crafted by one of Khodorkovskiy's closest associates, who left the wording sufficiently vague and subject to multiple interpretations, so that participation posed a high legal risk for foreigners because the investment could not be guaranteed.[52]

A second way in which Menatep manipulated the auctions for YUKOS was through the use of direct, personal intervention. For instance, senior representatives of YUKOS and Menatep appeared together at a joint press conference and on television a month before the auction to underline their intention of working together, in a further attempt to dissuade rival bids.[53] Furthermore, upon learning that his rivals had approached a wealthy foreign investor – Davis Petroleum Company – for a cash injection to make up the deposit for the investment auction, Khodorkovskiy sent one of his closest associates to meet with its owner who was advised that he was likely to lose his investment in view of the vague legal provisions governing YUKOS's auction, the underdeveloped judiciary system in Russia and the influence of Menatep and YUKOS in political circles.[54] Having neutralized the threat of indirect foreign participation, Khodorkovskiy attempted to do likewise with rival domestic bidders. In late November 1995, Khodorkovskiy reached a verbal agreement with Vinogradov, whereby in return for Menatep's sale of its shares in the confectionery factory Babaevskoe to Inkombank, the latter would concede the YUKOS auction to Menatep.[55]

However, the use of the above legal and personal tactics did not result in a smooth auction for YUKOS's shares. On 27 November 1995, the rival consortium of Inkombank, Al'fa bank and Rossiyskiy Kredit issued a joint statement which pointed out that with the YUKOS bid, Menatep's exposure would increase to 'a level beyond US$1.1 billion, which was more than ten times its own banking capital'.[56] It also claimed that Menatep's bid was being financed through 'the use of state resources in the Ministry of Finance'.[57] As such, it proposed that the auctions for YUKOS be postponed to modify some of its conditions. On 29 November, the Ministry of Finance issued a statement that rejected the said claim as 'unfounded'.[58] On 30 November, in response to the consortium's threat to liquidate government bonds in its possession if it could not use the bonds as partial payment of the deposit, the Central Bank warned of retaliation if the bond market was destabilized in this manner.[59] On 1 December, GKI issued a press release which addressed the consortium's concerns about the auction conditions, to which no changes would be made.[60]

On 5 December, the final day to register bids for YUKOS, Menatep bank, as the auction organizer, refused to accept the consortium's bid for the investment auction because the deposit was contrary to the auction rules; it comprised only US$82 million in cash and a document indicating possession of government bonds, the value of which could cover the rest of the deposit. The consortium's bid to participate in the trust auction was not accepted either, since the conditions stated that only participants in the investment auction could bid in the trust auction. At the auction on 8 December, there were only two bidders, both of which were YUKOS-affiliated companies. Their presence was nonetheless crucial, since according to the rules, a minimum of two participants was required for the auction to be legally valid. Menatep thus obtained control of 33% of YUKOS's shares at a price of US$150.1 million, as well as the right to manage the state's 45% stake in the company exchange for US$159 million worth of credits to the government.

Following the loans-for-shares auctions, Khodorkovskiy and his associates proceeded to establish Menatep's control over YUKOS in a variety of ways. The first step was to integrate YUKOS into the structure of Menatep's holding company, Rosprom, which was intended to strengthen its financial management and also make it 'as difficult as possible to reverse'[61] the result of the auctions. In this connection, Muravlenko was appointed as chairman of Rosprom (and retained his post as president of YUKOS) while Khodorkovskiy became the chairman and first vice-president of YUKOS while serving as president of Menatep. A number of Khodorkovskiy's closest associates in Menatep were likewise appointed to YUKOS's senior management.[62]

Second, Menatep established centralized control by YUKOS over its subsidiaries through a series of written agreements and personnel changes.[63] Henceforth, subsidiaries were no longer financially independent as all revenue and expenditure streams were centralized at YUKOS itself. Simultaneously, personnel changes were made to the management and boards of directors of rebellious subsidiaries. For instance, Tarkhov, the aforementioned general manager of NovoKuybyshev, was removed from his position and co-opted as one of the vice-presidents of YUKOS, thus securing his acquiescence with the new arrangement.

The third tactic that Menatep employed to consolidate its control over YUKOS, while acting as trustee of the government's stake, was to dilute the latter's shareholding through a recapitalization of the company. By the mid-1990s, many Russian enterprises had accumulated huge amounts of debt; YUKOS and its subsidiaries, for example, owed almost two trillion rubles to the federal budget in unpaid taxes, while the wage arrears for Yuganskneftegaz alone amounted to 350 billion rubles.[64] To reduce the scale of arrears, a decree was issued in July 1996 permitting enterprises with state shareholdings of 25% or more to issue new shares.[65] YUKOS duly held a subscription for additional shares and raised almost 500 billion rubles of extra capital. Consequently, the state's shareholding fell from 45% to 33.3% (the state did not purchase any extra shares on offer), while Menatep increased its stake to over 51% of YUKOS's shares. In this way, Menatep made use of legal provisions for debt restructuring 'as an efficient means of strengthening ... control in a corporation',[66] since the amount raised was far from adequate to completely pay off tax arrears.

The above measures were implemented with the expectation that upon the expiration of the pre-set term for trust management on 1 September 1996, Menatep, which as an auction winner had the right to sell these shares, would be in an advantageous position to then acquire these same shares. Indeed, in the collateral auction for 33.3% of the state's shares held on 23 December 1996, a front company backed by Menatep emerged as the winner with a bid of US$160.1 million, which was fractionally above the starting price of US$160 million. In a repeat of the previous year's auctions for YUKOS, Menatep again manipulated the auction conditions through its appointment as auction organizer. First, by declaring that the stake would be sold as a single lot, Menatep effectively barred foreign investors from participating, since the latter were not allowed to own more than 15% of an oil company. Secondly, by including the auction condition that the winner would have to invest an additional US$200 million over two years into YUKOS, Menatep ensured that few bidders would be interested in a financial commitment of US$360 million for a minority stake, especially since Menatep controlled more than 51% of shares. Thirdly, Menatep again ensured that the auction was valid by registering a second affiliate to bid in the auction. Finally, the starting price of US$160 million was specifically set to allow Menatep to recoup its initial investment of US$159 million made at the previous trust auction. Indeed, given that the commission agreement allocated to Menatep 30% of sale proceeds above the original collateral auction price, Menatep paid only an additional US$700,000 for the state's shares in YUKOS! In total, Menatep spent US$509.7 million to acquire YUKOS, which was later subsequently valued at US$1.4 billion in 1998, US$3.8 billion in 2000 and US$19 billion in 2002.[67] YUKOS therefore became the first privately owned oil company in Russia, with Menatep controlling almost 85% of its shares.

This case study of YUKOS has thus far considered the roles of Muravlenko and Khodorkovskiy in corporatizing and privatizing the oil company. It is now necessary to examine the role of a third actor in this process – the state. In the first place, the state played a crucial role in the initial corporatization of oil companies, as was noted earlier. It suffices to highlight the fact that the resulting decree 1403 was the outcome of mutually reinforcing factors. In this regard, pressure from the 'oil generals', who had *de facto* control over oil enterprises, was complemented by Mintopenergo's perception of VIOCs as a way to retain some state control over the oil industry. This was, in turn, in line with willingness by GKI and the Anti-Monopoly Committee to compromise on VIOCs and corporatization as the first steps towards a complete withdrawal of the state from ownership of oil companies.

Secondly, the state played a crucial role in the loans-for-shares scheme. In reshaping Potanin's initial proposal and subsequently approving the scheme, Yel'tsin and GKI were primarily motivated by political considerations, specifically to maximize Yel'tsin's re-election chances and, to a smaller extent, by budgetary concerns. In the case of YUKOS, Menatep appeared to have benefited from the largesse of its main client, the Ministry of Finance, which habitually took out large bank loans to cover temporary budgetary shortfalls and then paid the loans back with high interest. In fact, Menatep and ONEKSIMbank accounted for the lion's share of the US$1.32 billion earned by 'authorised' banks in this way between 1995 and 1996,[68] which in turn probably accounted for part of the cash that Menatep was able to raise for the YUKOS bid. Another way in which elements of the state facilitated Menatep's acquisition of YUKOS was illustrated by the behaviour of the auction commission chaired by GKI; it objected neither to Menatep being named as auction organizer nor the additional conditions that it imposed which allowed the bank to manipulate the auctions in its favour. Furthermore, as was earlier mentioned, a number of state institutions, including GKI, the Ministry of Finance and the Central Bank, also actively supported Menatep against accusations made by the rival consortium in the run-up to the auctions.

Third, various state institutions facilitated Menatep's consolidation of ownership over YUKOS, by putting pressure on the bank's rivals – the consortium comprising Inkombank, Al'fa bank and Rossiyskiy Kredit – to withdraw their legal suit aimed at annulling the results of YUKOS's collateral auctions and investment tender. The suit was dismissed in June 1996 but Inkombank was undeterred and appealed the decision. Around this time, rumours surfaced in the Russian media that Inkombank was suffering from a serious liquidity problem and was about to go into temporary administration. Subsequently, depositors made a run on the bank and several major banks suspended trade with it. These rumours were based on an unflattering draft audit report by the Central Bank, which Vinogradov claimed had been written 'on someone's orders'[69] in retaliation for Inkombank's law suit. Then in July 1996, the Central Bank issued a revised report expressing confidence in Inkombank, and soon after, the latter agreed to abide by the results of the YUKOS auctions.

On the one hand, it was possible that the Central Bank's audit of Inkombank was not linked to the latter's legal challenge, since by August 1996, five banks had been put under temporary Central Bank administration as a result of the introduction of more stringent charter capital requirements.[70] On the other hand, the Central Bank had threatened to retaliate against Menatep's rivals in the run-up to the loans-for-shares auctions, and moreover, media outlets with close ties to Menatep (*Kommersant'* newspaper was owned by SBS-Agro, which had provided some of the funding for Menatep's bid for YUKOS), and Berezovskiy (he was allowed to control ORT television even though the state owned 51% of shares) had been most active in covering the Inkombank saga. It is therefore difficult not to conclude that parts of the state bureaucracy, particularly GKI and the Central Bank, had worked together with the auction winners to pressure Inkombank to desist from undermining the YUKOS auctions and loans-for-shares in general.

The fourth way in which certain bureaucratic agencies influenced the privatization process of YUKOS, was by creating the impression that the collateralized shares could and would be redeemed, and thereafter returned to state-ownership. As mentioned earlier, this was partly achieved through the appointment, in January 1996, of Kazakov as the new GKI Chairman, in the wake of the furore over the loans-for-shares auctions of 1995. Kazakov's appointment and concessions on privatization were, in fact, tactical moves on the part of GKI, based on the recognition that 'preserving privatisation as a whole required concessions and trade-offs, at least until June 1996'.[71] GKI was so successful in creating the impression that the shares could and would be redeemed that many analysts at that time discounted the probability that the banks would put the stakes up for sale.[72] This more congenial reception to the controversial scheme in turn minimized the danger that Yel'tsin could be forced to abrogate it as part of a populist campaign measure. Menatep was thus able to use the 'breathing space' to consolidate its control over YUKOS.

Finally, the state also played a crucial role in the decision not to redeem the loans from the auction winners and instead to allow the latter to sell the collateralized shares after 1 September 1996. This decision was jointly adopted in late September 1996 by the government and the Kremlin's Security Council, which had been tasked by Yel'tsin to work together on this issue, and whose new members included Potanin as deputy prime minister and Chubays as the head of the Presidential Administration. As was the case for the earlier loans-for-shares scheme, political considerations provided the overriding impetus for the decision in three important respects. In the first place, for Chubays and Potanin, it was the logical conclusion to the earlier agreement to hand the tycoons the 'second key' to Russia's leading companies only after Yel'tsin's re-election.[73]

In addition, the decision was the result of a trade-off with Shafranik. The latter was keen to ensure that Mintopenergo (and his own position) was not made redundant by a privately owned oil industry and who therefore sought to retain as much state-ownership of the oil industry as possible.[74] In return for Mintopenergo's support to proceed with the sale of collateralized shares, GKI agreed to give up to Mintopenergo its prerogative to name state representatives to the board of directors of companies with federal shareholdings. Also, GKI and the Anti-Monopoly Committee agreed to jointly sponsor a decree proposed by Shafranik and signed by the President in August 1996 'On Prolonging the Period of Retaining as Federal Property the Shares of Joint-Stock Companies in the Fuel and Energy Complex'. Accordingly, state-owned shareholdings of strategic companies would be retained until 31 December 1998, and not divested earlier as had been planned.[75] These were shrewdly conceived concessions by GKI, based on the recognition that 'not only were private investors unable to pay for the privatisation of enterprises, they also did not have money to rescue these enterprises from crises'.[76]

The decision to proceed with the sale of collateralized shares was also made possible by an alliance to outmanoeuvre Aleksandr Lebed', the head of the Security Council. His solution to the share redemption issue was to sell these shares in lots of 10%–15% through an international tender to ensure maximum revenue for the state, and to enlist the help of the security organs to monitor and manage the auction conditions. Prime Minister Chernomyrdin, who regarded Lebed' as a potential rival for the Presidency, allied himself with Chubays and Potanin because the proposal by Lebed' would imply that 'the Security Council and no other organ, would be managing the collateralized shareholdings of the state',[77] thus raising its influence and profile at the expense of the government. In any case, Lebed' capitulated and as a result, Menatep became the first auction winner to sell the collateralized shareholding of the state in YUKOS.

The YUKOS affair

Khodorkovskiy spent the next few years attempting to consolidate and enhance YUKOS's stake in its various subsidiaries, which averaged 51% at the end of 1996.[78] By 2000, he had obtained near-absolute control of YUKOS and its subsidiaries, which gave him the incentive to increase, rather than destroy, the value of the company and hence his own shareholdings. YUKOS quickly became the darling of investment funds thanks to its self interested commitment to growth, profitability, cost efficiency and transparency. In 2002, it surpassed Gazprom as the most valuable listed Russian company even though the latter's sales were 30% higher, and in 2003, it replaced LUKoil as the industry leader in terms of revenue and production levels.

Things began to unravel for YUKOS in mid-2003. In June, a provocative report warned of an impending 'oligarchic coup' to remove the presidency in favour of a parliamentary republic with Khodorkovskiy as prime minister.[79] At the end of June, YUKOS's head of security, Aleksei Pichugin, was arrested on charges that he had organized a series of contract killings in 1998, and a few days later, Khodorkovskiy's long-time partner and chairman of Menatep, Platon Lebedev, was arrested for fraud dating back to 1993–1994. Thereafter, premises belonging to YUKOS and its subsidiaries were raided by the police searching for

evidence against Pichugin and Lebedev. In October, Khodorkovskiy himself was arrested at gun point and later sentenced to eight years in prison.
In the wake of claims by the Tax Ministry for back taxes amounting to
over US$30 billion, 76.8% of Yuganskneftegaz was sold for US$9.4 billion
in December 2004 through a court-mandated auction to a bidder that
was eventually acquired by state-owned Rosneft'.[80] Further auctions of
YUKOS's assets were held in 2007, all of which were won by Rosneft'. As
a result, YUKOS has effectively ceased to exist as an oil company, while
Rosneft' has emerged as the leading oil producer in Russia, accounting for
20% of the country's total, up from under 5% a few years ago.

Although privately owned companies still account for just over two-
thirds of Russia's oil output, this 'creeping nationalisation' is clearly a
reversal of policies aimed at creating a privatized Russian oil industry
since 1992. The following section will analyse the reasons for the
YUKOS affair and its implications on the relationship between business
and the state in policymaking.

There are three possible ways of interpreting the YUKOS affair.
The first, and least convincing, explanation is that advanced by the
prosecutor-general, tax authorities and Putin himself, according to which
'the YUKOS affair is connected primarily with taxes'.[81] Accordingly, the
intention of the state was 'not to go after specific individuals but to establish order in the country',[82] so as to 'demonstrate to all ... corporations
and selected individuals alike that taxes must be paid in their entirety'.[83]

One way in which YUKOS managed to minimize its tax obligations
was to take advantage of preferential tax rates accorded to its subsidiaries in impoverished regions within Russia that had been granted
the status of 'internal offshore zones'.[84] The fact that Evenkiya was
named an internal offshore zone was probably due to the efforts of its
representative in the Federation Council at that time, Boris Zolotarev, a
former YUKOS executive. In addition, YUKOS managed to repeatedly
delay the imposition of additional taxes on the oil industry. One of its
core shareholders, Vladimir Dubov, was the deputy chairman of the
Duma Committee on Budget and Taxes, and he actively defended the
company's interests in this respect.[85] Indeed, there was a perception that
'the oil lobby, led by YUKOS, was the strongest lobby in the Duma'.[86]
Another method of tax minimization concerned the practice of 'transfer
pricing', whereby a company's oil-producing subsidiary sells oil cheaply
to a trading subsidiary (usually located in tax havens) within the same
group.[87] This results in a lower tax burden on the less profitable oil-
producing subsidiary. Independent oil analysts have estimated that this
practice was used by many oil companies to lower their oil tax per ton

by an average of $10 in 1999 and by $13 in 2000.[88] Such schemes allowed YUKOS to enjoy an effective tax rate of 20% which was below the statutory corporate tax rate of 24%. In comparison, Surgutneftegaz paid the same amount in taxes as YUKOS even though its revenues were 40% lower, thereby implying an effective tax rate of 30% for Surgutneftegaz.[89]

The argument that the YUKOS affair was the result of the company's tax evasion practices is dubious for several reasons. In the first place, while there is no doubt that YUKOS made use of offshore accounts and loopholes within the tax code to reduce its tax burden, it was not the only, or even the worst, offender in this respect. In the case of Sibneft', for example, its effective tax rate for 2000 was merely 16%, and in other years it was as low as 12%.[90] This was because Sibneft' made extensive use of tax privileges within the internal offshore zone in Chukotka, whose governor Roman Abramovich was, until 2005, the largest shareholder of Sibneft'. In addition, Sibneft' staffed its trading subsidiaries with handicapped personnel in order to qualify for a 50% reduction in profit taxes.[91] In other words, 'transfer pricing ... was used by absolutely all the oil companies, and it was not considered a crime by any one, including the tax authorities'[92] until the YUKOS affair.

Second, tax minimization schemes were, in fact, perfectly legal. Putin himself noted the 'failings in the Russian legislation' that had allowed major corporations to take advantage of tax loopholes in the regions.[93]

Third, more than 70 attempts by YUKOS to settle the claim for back taxes went unanswered.[94] For example, the Tax Ministry refused to consider proposals to unfreeze the 44% of YUKOS shares held by key shareholders, which would have allowed them to dispose of their shares to raise money for the tax debts. It also failed to respond to a bail-out plan by a group of investors to pay off the company's tax debts and buyout key shareholders, including Khodorkovskiy. Moreover, it was not willing to discuss a phased tax repayment schedule. In other words, the state did not appear eager to recover back taxes *per se*. It was prepared to countenance only one solution, that is, the forced sale of Yuganskneftegaz and the effective destruction of YUKOS. This was because the tax claims were merely a means to a larger end.

An alternative way of explaining the YUKOS affair is to link the selective assault on the company and Khodorkovskiy to a 'redistribution of property, that is, the expropriation of YUKOS in favour of new owners' closely allied with Putin.[95] In this regard, the new elites within the state bureaucracy and presidential administration were no longer satisfied with political power; instead, they wanted to have both power *and* property. They comprise those lawyers and economists who previously worked with Putin in St Petersburg (the petersburgers or liberals) and representatives from the law enforcement and security agencies (the *siloviki*).[96]

There is ample evidence to support the 'property distribution' thesis. For instance, Vladimir Ustinov, the prosecutor-general in charge of the affair, is allegedly a leading member of the *siloviki*.[97] His prosecution of YUKOS was facilitated by an accusation brought against YUKOS by Sergey Bogdanchikov, president of Rosneft', who is closely allied with the *siloviki*, over a minor business dispute; little wonder then that Bogdanchikov and Rosneft' have been the biggest beneficiaries of YUKOS's demise.[98] Furthermore, a number of high-ranking *siloviki* have been appointed to the boards of directors of state-owned companies in the airline, defense and energy sectors. For instance, Igor' Sechin, the 'spiritual father' of the *siloviki* and deputy chief of the presidential administration, was made chairman of Rosneft'; he was widely credited with masterminding the campaign against YUKOS.[99] Previously, only senior government bureaucrats were appointed in these positions; as such, the new appointments policy transferred control of property belonging to the state in general to a particular, narrow segment of the bureaucracy. For the *siloviki*, property redistribution – via transfer of assets (in the case of YUKOS) and of *de facto* ownership rights (as per the appointments in state-owned companies) – serves as an avenue for self-enrichment and may also ensure that, as the Yel'tsin-era tycoons, they will have considerable political and economic weight to select an ally as the next president in 2008.

The 'property redistribution' thesis of the YUKOS affair is useful because unlike the focus on fraud and tax evasion, it provides a convincing explanation as to why YUKOS was singled out, by referring to its commercial disputes with Rosneft', a company allied with the *siloviki*. However, it is at best a partial explanation since it tends to portray Putin almost as a bystander in a business dispute between YUKOS and Rosneft'. It thus ignores the question of why Putin, whose highest-order preference is his political longevity, would allow Russia's stability and investment climate as well as his own political future to be put at risk because of commercial rivalry.

The final interpretation of the YUKOS affair posits that it was simply about politics and power, namely, that Khodorkovskiy's increasingly independent behaviour and business policies undermined Putin's ambitions for Russia as an energy superpower. For Putin, the enfeebled state of Russia's military meant that non-traditional sources of influence have to be sought if the country was to remain a great power.[100] In this regard, mineral resources, and oil and gas in particular, have been identified as intrinsic to the country's economic growth and international prestige, not least by Putin himself; he has argued that the development of the resource sector cannot be left entirely to market forces but must be a joint effort with the state.[101]

Since Putin's legitimacy and power as president appear to be directly dependent on his efforts to establish Russia as a leading actor in international oil politics, control over a crucial vehicle of influence in the global arena today – the energy sector – is imperative, thanks to soaring demand for oil in recent years.[102] As a result, Putin has sanctioned the creation and rise of the so-called Kremlinneftegaz, comprising Gazprom (with Sibneft' since end 2005) and Rosneft' (including most of YUKOS).

However, Khodorkovskiy's aggressive business strategies to expand YUKOS's international profile challenged Putin's control over a resource-based foreign policy. This was manifested especially in Khodorkovskiy's enthusiasm to build privately owned pipelines.[103] Since every 1% increase in exports adds about US$500 million to an oil company's net market value,[104] thereby increasing the value of shares held by top management, the bottlenecks in the state-owned pipeline system became a source of frustration for Khodorkovskiy and owners of other oil companies. Nevertheless he was much more vocal and abrasive in his criticisms than most of his peers.[105]

In this connection, YUKOS proposed the construction of a privately financed oil pipeline in the Far East from Angarsk (in Russia's East Siberia) to Daqing (in China) in order to supply the energy needs of a rapidly modernizing country. In October 2002, YUKOS, together with Transneft', submitted a proposal to the government to amend the legislative basis to give private investors the incentive to finance oil pipelines, for instance, by linking pipeline export quotas to investment levels.[106] YUKOS envisaged the acquisition of a 50% stake in the Angarsk–Daqing project and of half the pipeline's export capacity, so that by 2005, YUKOS would account for one-quarter of China's oil imports.[107] Then in November 2002, four leading oil majors – YUKOS, LUKoil, TNK and Sibneft' – signed a memorandum of understanding to finance, build and own a new pipeline from west Siberia to the ice-free port at Murmansk, from where crude oil would be shipped to the United States. As the key sponsors of the plan, YUKOS and LUKoil would each have a one-third stake in the project.[108]

For Putin, state-ownership of the pipeline network, along with oil-related taxes, is one of the few remaining levers that allow him some measure of control over oil companies, particularly privately owned ones. YUKOS was clearly not the only company attempting to reduce the extent of control exercised by the state over its oil exports and profits. Nevertheless YUKOS was singled out because it was perceived to be the key instigator. Moreover, even the smallest concession to potential investors in pipeline projects 'would set a huge precedent. The government would then not be able to ensure that it keeps full ownership of the electricity grid and gas pipeline network'.[109]

Another manifestation of the conflict between Khodorkovskiy's business strategies and Putin's energy-based foreign policy concerned the best way to supply oil to the fast growing, but energy-poor markets of East Asia. YUKOS eventually agreed that Transneft' would build and own the pipeline, but continued to champion the Angarsk–Daqing route against the alternative Angarsk to Nakhodka (in Russia's Far East) option. This was because the shorter distance to Daqing meant that the pipeline could be completed much more quickly and cheaply than the alternative route, thereby allowing YUKOS to take greater advantage of the high oil prices.[110] Transneft', however, backed the Nakhodka option because of wider socio-economic and geopolitical benefits for the country. Russian oil would be able to access various markets from Nakhodka, whereas the single destination Daqing pipeline would hold Russia hostage to energy demand and any political upheaval in China. Also, a pipeline going all the way through East Siberia, instead of being diverted midway to Daqing, would have a positive impact on the less developed regional economies there.[111] To the extent that the issue of the Far Eastern pipeline reflects a 'political choice defining Russia's strategy in northeast Asia',[112] Khodorkovskiy's interference was an unacceptable challenge to Putin's personal authority in foreign policy. He also undermined Putin's belief in the state as the 'main driving force of any change'.[113]

Finally, Khodorkovskiy's intention to sell 25–40% of YUKOS to either ExxonMobil or ChevronTexaco also brought it into conflict with Putin's desire to control the oil industry.[114] Putin was, in fact, not categorically opposed to partial foreign ownership of Russian oil companies, as evidenced by the acquisition of a 50% stake in TNK by British Petroleum in February 2003. The difference was that whereas Al'fa group, the parent company of TNK, possessed an abundant stock of relational capital vis-à-vis Putin – it is a relatively reliable and loyal ally of the state, with former executives serving as senior officials within the presidential administration – YUKOS did not.[115] In this regard, the 'involvement of foreign capital in Russia's most powerful oil company would diminish the Kremlin's ability to control it'.[116]

To recap, the singling out of YUKOS was connected with, above all, the politics of oil and the legitimacy and power that Putin derives as leader of an energy superpower; in order to have a significant influence over the political succession process, it was vital for him and his supporters to exercise control over the country's largest oil and gas companies. This is because 'oil presents similar possibilities as nuclear weapons did some time ago ... Thanks to oil, [the] President is accepted into any club ... Oil allows him to act as the leader of a superpower which has something the world depends on'.[117] Putin could not afford to allow the country's largest oil producer to be run by a tycoon who was perceived as becoming 'too big, and increasingly independent, a political and economic player'.[118] If YUKOS had been an equally successful chain of supermarkets or a real-estate developer, *ceteris paribus*, the YUKOS affair would not have happened.

Observations from the case study

Primacy of individual actors and their choices

The case study of YUKOS underscores the centrality of agency-based explanations of change. Muravlenko and Zenkin did not set out to reform the Soviet oil industry – this role belonged to Alekperov – but they did make the decision to emulate Alekperov and to use their influence within the government to create one of the first three corporatized oil companies in Russia in 1993. Policy emulation was, in fact, not a foregone or automatic behaviour, since they could have adopted a passive, non-committal approach as was the case with the general managers of other large oil production entities such as Nizhnevartovsk and Noyarbsk, which became part of holding companies TNK and Sibneft', respectively, only in 1995.

The extent to which choices by individuals affected outcome was likewise demonstrated by state bureaucracies in those formative years. As was mentioned earlier, at the end of 1991, the former Soviet Ministry of Oil chose to transform itself into Rosneftegaz and quickly lost its relevance. Its leader Churilov admitted that the wrong decision had been made and acknowledged that 'we just could not bring ourselves to break away from state control. Perhaps it was the inertia inherited from our Soviet past that made us cling to the centre'.[119] In contrast, however, this same 'Soviet past' galvanized GKI and Mintopenergo into embracing a second alternative, and they emerged as key players in formulating the legal basis for corporatizing and privatizing the oil industry.

It is certainly true that path dependency does not deny the possibility of change, which can take place only rarely at critical junctures, after which the choice that is made is locked in and determines the path of development henceforth. For the sake of argument, let us accept the contention that the critical juncture of the end of 1991 and the beginning of 1992 gave rise to new political leaders who, working together with the pre-existing economic *nomenklatura*, chose to de-monopolize and corporatize the oil sector with a view to privatization, and that, as such, the privatization of YUKOS may be considered a natural outcome of the initial decision.

This perspective, however, ignores the many challenges and obstacles between 1993 and 1996 which could have retarded or halted the process of privatization in YUKOS by Menatep. These included the aforementioned rival bid for YUKOS and the subsequent legal action to annul the auction results, pressure from across the political spectrum to redeem the collateralized YUKOS shares and Shafranik's plan to extend the period of federal shareholding in all strategic companies.

Even Muravlenko's agreement with Khodorkovskiy was not an ironclad guarantee that Menatep would succeed in acquiring YUKOS. This was because of tensions in the relationship between the two men as a result of Khodorkovskiy's attempts to wrest from Muravlenko those powers which the latter held as chairman, as well as Khodorkovskiy's cancellation of many social projects YUKOS had previously financed in Nefteyugansk. Indeed, 'a lot of effort had to be spent so that the relationship did not collapse completely'.[120] There was nothing 'natural' or predetermined about YUKOS's corporatization and privatization; interested parties had to make a series of choices to ensure the successful privatization of YUKOS by Menatep.

The role of institutions: Legacies and incentives

Legacies and initial conditions were important to the extent that they formed part of the informal institutional parameters – constraints and inducements – within which decisions were considered and adopted. Nevertheless in most cases, they did not predetermine an actor's behaviour, particularly when faced with a new set of incentives that appealed to his self-interest. Muravlenko was certainly more conscious of and beholden to historical considerations – such as work practices, management style and social commitments that characterized the Soviet oil industry – than Khodorkovskiy. Although asset-stripping may have been a consequence of the disarray of the Soviet period, Muravlenko was nevertheless able to respond to material incentives offered in the post-Soviet period; his growing contacts with investors and their interest to acquire his company. In doing so, he amended his strategy for self-enrichment from that of a 'roving bandit' to a 'stationary bandit' and advanced decisively the course of privatization in YUKOS.

In a similar way, the informal agreement of July 2000 between Putin and the business tycoons not to mix business with politics, failed to regulate behaviour in the YUKOS affair. As convincingly argued by William Tompson, both Putin and Khodorkovskiy developed a 'commitment problem' within a few years because the conditions under which the original bargain were based on became less applicable.[121] For Putin, his success at consolidating power by late 2003 presented him with an enormous incentive to renege on the terms of the 'contract' he had agreed to when his hold on power had been much weaker. For Khodorkovskiy, the carefully cultivated relationships with influential Western personalities after 2001, including Jacob Rothschild, Dick Cheney and Lord Owen, seemed to provide him with a '*krysha*' (literally 'roof' or protection) to oppose government policies deemed to impinge on his business strategy.[122] The resulting 'defection' from the July 2000 status quo reflects again the centrality of agency-centred explanations – in this case, the motivations behind Putin's and Khodorkovskiy's behaviours – in analyzing institutional change.

What of the role of formal institutions? In Putin's case, his career with one of the foremost Soviet state institutions, the KGB, and his internalization of its institutional norms and rules reinforced his behaviour with regards to the YUKOS affair. For example, he shared the high premium it placed on control and predictability and hence did not appreciate Khodorkovskiy's independence, particularly in foreign economic policy. The KGB experience also strengthened Putin's belief in the merits of a supercentralized state in Russia, a characteristic 'laid down in its genetic code, its traditions, and the mentality of its people'.[123] Therefore, he was not prepared to countenance Khodorkovskiy's repeated criticisms of and rivalry with state institutions, and his attempts to undermine the prerogative of the state in foreign policy. Nevertheless the institutional legacies of Putin's KGB career do not adequately explain his assault on YUKOS – they merely provided him with the tools and ideational basis with which to act against the company and leader. It was Putin who decided that energy, even more than military might, would form the basis of a strong state; this was arguably a self-interested choice since Russia's role as an energy superpower would greatly enhance his own legitimacy, authority and political future at home.

Relationship between the state and business:
The role of self-interest

Finally, the case study of YUKOS suggests that the nature of the relationship among the state, tycoons and *nomenklatura* managers of oil companies has a decisive influence on the privatization (or de-privatization) of oil companies in Russia. Khodorkovskiy's success was dependent on support from the *nomenklatura* manager of YUKOS, who had laid the groundwork for corporatization, and on backing from parts of the state bureaucracy. Unanimous support from all three actors in terms of the buyer, conditions and timing of the sale appeared to be crucial. YUKOS's management under Muravlenko publicly backed and approved Menatep's bid before the auctions and even during the period of trust management, because they wanted to enhance the long-term value of their own shareholdings in a privatized and better-managed company.

On the part of the state bureaucracy, support from GKI appeared to be particularly significant in facilitating Menatep's acquisition of YUKOS. As noted earlier, GKI's support for privatizing oil companies stemmed from the fact that privatization in general was not only its *raison d'être*, but was also an integral part of its bureaucratic competition with Mintopenergo to acquire oversight of the strategic and lucrative oil industry. GKI's pre-eminent position within the state bureaucracy rested, first of all, on formal legislation. It was, after all, the lead agency responsible for privatization, and its chairman also served concurrently as deputy prime minister in the government. GKI's powers included managing the state's shareholdings, making recommendations on the sale of shares and issuing voting instructions for state representatives in corporatized entities. Indeed, 'the legislation basis of privatization was such that the final voice was that of GKI'.[124] In comparison, the government or cabinet was responsible for approving or rejecting GKI's recommendations on sale of shares and the list of state representatives proposed by various bureaucracies. Mintopenergo's role in privatization was limited to proposing candidates to act as state representatives and to participating in the decision-making process on the transfer, sale and disposal otherwise of the state's shareholdings. As for the Russian Federal Property Fund [henceforth RFFI], it only implemented and organized the actual sale of state shareholdings.[125] Even though privatization schemes in the oil industry had to be approved by various bureaucracies, it seemed that 'all signatures [on the documents] except for GKI, were formal'[126] and 'nothing could happen without GKI's support'.[127]

The above-mentioned collusion between the state and business likewise facilitated the de-privatization of Yuganskneftegaz. The leading role played by GKI in privatizing YUKOS was, in the case of the YUKOS affair, taken over by representatives and allies of the *siloviki* within the bureaucracy, the presidential administration and the business community. The tycoons from the private sector also participated in the demise of YUKOS: not only were they too concerned about the fates of their own business empires to speak up in YUKOS's defence,[128] they also blamed Khodorkovskiy for provoking the authorities to act against the company. Chubays, for instance, noted that Khodorkovskiy 'started playing at politics, but he didn't play by the rules. He undertook to buy up the Duma. In their response, the authorities also broke a few rules'.[129] Hence, the self-interested motivations of state agents and leading tycoons from the private sector reinforced each other and made possible the de-privatization of Yuganskneftegaz.

Finally, what of the role of presidents Yel'tsin and Putin in the history of YUKOS? It should be clear from the case study that both presidents were primarily motivated by the same, self-regarding desire to stay in office and consolidate one's power through the creation of a coalition of supporters. In this connection, Yel'tsin approved the privatization of state assets mainly because it created a new base of support for him, especially among the tycoons, thereby reducing the chances of a communist victory during presidential elections. This is not to deny that Yel'tsin genuinely believed in the superior efficiency of a market economy; indeed, he has described how he became increasingly annoyed with the rigidities and failures of the command economy.[130] Nevertheless his highest-order preference was for power and much of his behaviour has been predicated on an intention to maximize this interest.

As for Putin, he was in favour of returning YUKOS's key assets to state control. He correctly calculated that this decision would greatly increase his authority and legitimacy at home as the leader of an energy superpower, as well as reward or reinforce his coalition of supporters, thereby enhancing his own political longevity. It is not the intention of the book to ignore the role that Putin's beliefs and experiences – as exemplified in his dissertation and KGB career – have played in shaping his behaviour in the YUKOS affair.[131] Nevertheless as Chapter 5 will explain, Putin was motivated more by an interest in his political tenure and longevity, rather than any ideational goal of strengthening the state; this is because Putin's political and economic policies were deliberately aimed at strengthening the presidency at the expense of the state.

4
Case Study of Slavneft'

Following Putin's victory in the presidential elections of March 2000, he defined a new modus operandi for relations between the state and big business. Accordingly, in exchange for the withdrawal from politics of big business, the latter could retain property that had been acquired through questionable means. Representatives of big business would no longer receive any preferential treatment from the state, and their interactions with state officials would take place through formal, institutionalized channels instead of through biased, personal relationships. In this regard, it was expected that privatization under Putin would be qualitatively different from that under Yel'tsin. Indeed, the first major privatization under Putin, the sale of ONAKO oil company in September 2000, was hailed as the 'best privatisation Russia ... had'[1] until then, given that it was 'honest, open and absolutely competitive'[2] and based on 'a competition between financial proposals'[3] instead of political connections.

This chapter argues that while the mechanics of corporatizing and privatizing Slavneft' were rather distinct from that of YUKOS, the outcome was the same, since the winning bid by Sibneft'–TNK was manipulated and predetermined long before the actual sale held on 18 December 2002. As in the case of YUKOS, the dynamics of the triangular relationship between the company's internal management, potential buyers and the leading representatives of the state determined the scope and pace of privatization. The transfer of power from Yel'tsin to Putin, and the new framework of business–state relations outlined by the latter, therefore appeared to exercise limited influence on the privatization of Slavneft'.

The first part of this chapter will study the creation and corporatization of Slavneft' up to 2000, in order to analyse the motivation and interests of those actors who influenced the process of the company's initial but limited privatization and to account for the failure during this period to sell the rest of the company. The second part of this chapter will focus on the period from 2000 to 2002, explaining how and why the perspectives of these same interested parties changed in favour of selling Slavneft'.

Creation of Slavneft'

It has become almost a truism to suggest that in Russia, the corrupting influences of 'personal ties and personal interests of bureaucrats were the sole criterion'[4] for decision-making. In this regard, a senior oil executive noted that with respect to creating oil companies, 'what influenced decision-making were ... certain private decisions'.[5] The creation of Slavneft', however, was the exception to this rule: in this case, personal relationships certainly did play a role, but historical, economic and political considerations were equally important.

By the end of 1991, the leading *nomenklatura* general managers had largely accepted the creation of VIOCs as the panacea for the problems facing the Soviet oil industry. LUKoil was in discussions to merge with suitable refineries while Kuybyshev refinery had just approached Yuganskneftegaz with this aim in mind. At that time, Anatoliy Fomin was one of the informal leaders of the Russian oil industry:[6] he was the general manager of Megionneftegaz, an oil production association located within the oil-rich Tyumen' oblast' that accounted for 3.6% of the oil output of the Russian Federation in 1992.[7] He was also the co-founder of Bank Yugra, which handled Megionneftegaz's financial transactions. In 1991, he visited refineries in Belarus and Czechoslovakia to evaluate the feasibility of a merger with Megionneftegaz. Unlike LUKoil or YUKOS, his plan was to enter into a partnership with a refinery that was not located within Russia but which was as close as possible to the rich, consumer markets in Western Europe in order to minimize transportation costs.[8] By mid-1991, he reached a deal with a Czech refinery but the agreement was never implemented because of political upheavals and government changes in first, the Soviet Union, and then, in Czechoslovakia. In April 1993, Fomin was appointed deputy minister of Mintopenergo and was succeeded at Megionneftegaz by his partner, Anatoliy Kuz'min, with another friend Sergey Alafinov serving as director of finance. The company quickly became an active exporter of oil, thanks to what one newspaper referred to as the 'special goodwill' of Mintopenergo.[9] This was because as deputy minister, Fomin was responsible for allocating monthly export quotas for crude oil which were exempt from taxes. In return, he received a share of the proceeds of Megionneftegaz's oil exports to the world market through a foreign intermediary, which was itself owned by him, his partners at Megionneftegaz and some foreigners.[10] Given that the domestic price of oil at that time was less than 5% of the world market price, the rents from such transactions were enormous.

The opportunity to revive the attempt to create a transnational VIOC centered on Megionneftegaz, and hence to stake a claim to part of the post-Soviet oil industry, soon presented itself. As will be argued below, there was a shift in the conception of Russia's foreign policy interests in its former republics which helped to develop a common ideas-based cognitive framework. This gave rise to a consensus within the foreign policy elites regarding the interpretation of Russia's interests and thereby narrowed the range of available policy options. Fomin took advantage of this ideational consensus and used it to appeal not only to the divergent interests of decision-makers in Russia – including Mintopenergo, GKI, the president and the government led by Chernomyrdin – but also those in Belarus, most notably its Prime Minister Vyacheslav Kebich.

Ideas, according to Jeremy Bentham, are 'talk ... [that] do not help us to understand. Rather they obstruct our vision'.[11] Likewise, Kenneth Shepsle has opposed the role of ideas in explaining an individual's goal-motivated behaviour and has opined that ideas are mere 'hooks on which politicians hang their objectives and by which they further their interests'.[12] Since the late 1980s, however, there has been a noticeable trend in academia towards bringing ideas back in to political analysis.[13] One approach of reconciling behaviour and ideas in politics has been to assign causal primacy to ideas or beliefs, so that they, rather than interests, explain the behaviour of political actors. For instance, as noted in Chapter 2, the ideas espoused by epistemic communities are sometimes said to account for the global spread of privatization. In Russia, it has also been argued that the refusal by liberal politicians to unite into a single party is due to their divergent interpretations of market-oriented reforms and democracy, rather than a competition for power.[14]

An alternative approach to conceptualizing the interplay between interests and ideas is to view the latter as focal points or points of reference for collective action by rational actors keen to maximize or 'satisfice' their preferences.[15] In this respect, ideas mitigate differences between actors, allow compromises to be struck and strengthen coalitions. They function in much the same way as informal institutions by constraining, reinforcing or expanding an actor's policy options in a given situation: they are a common denominator in encouraging cooperation and they delimit the decision-making context within which self-interested actors conduct policy discussions and attempt to influence policy outcome.

Fomin's creation of Slavneft' for personal gain succeeded because it coincided with and leveraged upon the rise of so-called pragmatic nationalism in Russia towards the end of 1992, supporters of which were in favour of a more activist and even integrationist approach in relations with its former Soviet republics.[16] Consequently, this new foreign policy orientation influenced much of Russia's actual behaviour in the CIS as exemplified by the country's participation in domestic intrigues in Georgia and Azerbaijan, in seeking joint-development of resources in the Caspian, in defending the rights of its Russian citizens living in the CIS and in carrying out less than even-handed peacekeeping operations in Tajikistan, Moldova and Georgia.[17]

Russia's main economic interest in Belarus was to seek repayment of energy debts. Prior to 1989, Russia supplied 95% of Belarus's oil and synthetic rubber and all of its natural and liquefied gas, while Belarus was assigned a 'finishing role' for these raw materials, in line with the Soviet policy of industrial specialization. Indeed, of the 120 million tons of Russian oil that was sent for refining in other Soviet republics in an average year during the 1980s, 80% was processed by refineries in Ukraine and Belarus.[18] The Druzhba oil pipeline in Belarus would then transport the refined oil westwards to European markets. Belarus was thus adversely affected when, as part of price liberalization reforms adopted in 1992, Russia began to charge world market prices for energy that had previously been supplied at subsidized prices. The cost of energy increased by nearly threefold, and by autumn 1994, Belarus owed Russia US$550 million in unpaid energy bills.[19] Faced with the reality of its vulnerability dependence on Russia – Belarus was unable to reduce its reliance on Russia by switching to alternative energy suppliers given the lack of hard-currency resources – it resorted to swapping debt-for-property.

The first debt-for-property swap was concluded in September 1993. According to the agreement signed by Chernomyrdin and Kebich, the state-owned gas supply company in Belarus, Beltransgaz, was incorporated as a subsidiary into Russia's Gazprom. Subsequently, in December 1993, the two prime ministers agreed to create joint-stock companies in the oil sector involving Belarussian oil refineries and Russian oil producers.[20] By this time, Megionneftegaz was delivering oil to the Mozyr and Novopolotsk refineries in Belarus, but Fomin made known his preference for a merger with Mozyr. The leadership was not only more professional – they consistently made timely payments for fuel deliveries – but they already had a plan in place for upgrading the capacity of the refinery in expectation of future expansion plans. Also, Mozyr was well located for export purposes: it was situated at the nearest point to Western markets just on the border of the former Soviet Union and along the main Druzhba oil pipeline to the west.[21] In April 1994, Chernomyrdin duly ordered GKI and Mintopenergo to conduct talks with Belarus about the creation of Slavneft' on the basis of two Russian companies, Megionneftegaz and Megionneftegazgeologia, and the Mozyr refinery in Belarus. Following months of negotiations, Slavneft' was officially incorporated as a Russian–Belarussian oil company in August 1994: its shareholders were the government of Russia (with 83.2% of shares), the government of Belarus (8.83%) and Mozyr refinery (7.97%). It therefore became the first oil company in Russia with foreign partners.

As part of the debt-for-asset swap, Slavneft' was given 17% of the shares in Mozyr refinery. Aleksandr Lukashenko, who had defeated Kebich in the bitterly contested presidential elections in July 1994, made no attempt to review the terms of the merger, and he facilitated cooperation by appointing a close associate of Fomin's as Belarus's deputy prime minister in charge of energy and industry, in an acknowledgement that 'there [was] no exit from our economic crisis without Russia'.[22] For the leaders of Belarus, the deal on Slavneft' ensured continued oil supplies to the impoverished republic, thus minimizing social and economic upheavals and enhancing their political longevity.

For the Russian government, the creation of Slavneft' was a useful economic lever to retain and extend its influence in Belarus. The fact that other inroads were made in Belarus by Russian companies suggests that Slavneft' was not merely a one-off project. More specifically, in late 1995, YUKOS and LUKoil jointly reached preliminary agreement with Belarus to establish a joint venture with 74:26 share ownership, respectively, to reconstruct the country's other main refinery, Novopolotsk. The two Russian companies also agreed on another joint venture with Belarus to build a new oil processing complex at Novopolotsk, 51% of which would belong to the Russian partners.[23] As for Slavneft', by mid-1996 it concluded further swap deals with Belarus, according to which its stake in Mozyr was increased to 42.6%, in return for an extension of credits to the Belarussian government to upgrade the refinery.[24]

The creation of Slavneft' by Fomin was thus the result of a confluence of choice and circumstance. Political changes in the Soviet Union and

Czechoslovakia forced Fomin to look elsewhere for a suitable refinery as partner, while Russia's new focus on the CIS narrowed the range of possibilities where the refinery could be found. In a revealing admission of how these elements came together, Dmitriy Romanov, the state representative from GKI on the board of Slavneft', noted that the founders of Slavneft' attempted to find 'a political and public relations niche with regard to economic co-operation with Belarus ... This was the Yel'tsin period from 1993 when in the sphere of public politics, friendship with Belarus was on the level of brotherly love and this Belarussian political factor was used for the benefit of the company'.[25] Following the murder of Kuz'min in mid-1994,[26] Fomin was appointed to head the newly constituted Slavneft', where he remained till 1998.

Failure of privatization: 1997–1998

1997 started as a good year for Russia in terms of progress in economic reform.[27] The political climate appeared to once again favour an intensification of market-style reforms, rationalization of the tax system and continuing divestment of state assets. Following a heart bypass operation, a reinvigorated Yel'tsin appointed Chubays and Nemtsov as first deputy prime ministers in March, with additional portfolios as Finance Minister and Fuel and Energy Minister, respectively. Other members of Chubays's team were also appointed to influential positions, including Al'fred Kokh as deputy prime minister and head of GKI, and Pyotr Mostovoy as Chairman of the Federal Bankruptcy Agency. Overdue pensions were paid by the summer thanks to success in forcing Gazprom to pay off most of its huge tax arrears and the receipt of a record US$1.875 billion for the privatization of 25% of telecommunications company Svyazinvest.

By this time, the Russian government had approved a privatization programme for its 83.2% stake in Slavneft': 8% would be sold at specialized auctions, 5% would be retained by an employee's fund, 20% would be sold at an investment tender and the remaining 50% would be held by the state.[28] During 1996 and 1997, multiple packages amounting to a total of 8% were auctioned off, mostly to small individual shareholders with the aim of raising money for the upgrading of Mozyr refinery. This was followed by an announcement by the RFFI in November 1997 regarding a commercial tender for a 19.68% stake in Slavneft', but the auction failed to attract any bidders. An attempt to organize a new auction for the same shares in March 1998 was abrogated. The December 2002 sale of Slavneft' therefore represented the third major attempt by the state to privatize the company.

One explanation as to why the 1997–1998 auctions for Slavneft' failed concerns manipulation by Russian tycoons who were unwilling to pay the high asking price and who, 'with the help of lobbying in the government and machinations with the law suit in the arbitration courts, contrived to twice annul the sale of Slavneft".[29] For example, in early 1998, a suit was filed in a local court by a worker, from Ryazan refinery, who alleged that the sale of Slavneft' scheduled for March 1998 was illegal because the advertisement for the sale and various tender documents were invalid. The court ordered the tender to be ceased pending investigations. Although it subsequently sanctioned the sale, the tender deadline had expired by then, and RFFI officially declared the tender invalid for lack of bids.

It appeared that the Al'fa group had engineered the failed tender – indeed, it controlled the majority of shares in Ryazan refinery – because it was interested in obtaining Slavneft' to complement operations in its other oil company, TNK, but was at that time unable to raise money for the tender.[30] As a result, TNK 'did everything just to slow down the whole process'.[31] Al'fa was, in fact, financially constrained: although a further 49% of TNK's shares were on offer in March 1998, Al'fa, which by that time already owned 40% of TNK, decided to purchase only enough shares to give it majority control of the company. In fact, Al'fa had resorted to this same tactic of relying on a court order to suspend auctions in November 1997 with the aim of buying more time to raise the money needed in an auction for 49% of TNK.[32] The subsequent attempts by TNK to purchase controlling packets of shares in subsidiaries belonging to Slavneft', and TNK's eventual purchase of Slavneft', in 2002, lends credibility to the explanation that Al'fa played an active part in getting the tender invalidated.

A second possible explanation for the unsuccessful 1997–1998 auctions for Slavneft' was the failure by the government, and RFFI in particular, to market and manage the sale properly.[33] In this respect, the stake of 19.68% was too small to be of interest to any potential owner as it did not even amount to a blocking stake and an automatic seat on the board of directors, where a 25% plus one share is required under Russian law. The lack of shares of Slavneft' on the secondary market also meant that it was not possible to purchase them with a view to increasing the stake on offer. In addition, Fomin's authoritative and combative style of management made prospective buyers even more wary of entering into a potentially fractious relationship with only a minority stake.[34]

Taking into consideration auctions of oil companies actually conducted between 1997 and mid-1998 (before the default in August), it is notable that successful outcomes were limited to those cases where the stakes on offer bestowed on the buyer nearly majority control – such as TNK (40%) and Vostochnaya Neftyanaya Kompaniya or VNK (45%). Those that failed, namely, Slavneft' and Norsi oil (14.5%) had much smaller stakes on offer.[35] Moreover, offering these companies for sale at the same time actually depressed their value, since the target group of buyers comprised only a small group of Western and Russian investors. Compounding this problem was the announcement that the highly prized Rosneft', at that time the fifth-largest Russian oil company, would be sold in mid-1998. As a result, buyers preferred to save their cash reserves in anticipation of the Rosneft' sale: this meant that less money was available to place bids in auctions for other oil companies.

In all probability, the government had little choice but to attempt to divest as many shares as possible even in the face of worsening market conditions – the Russian stock market fell 51% between October 1997 and January 1998 as investors were withdrawing from perceived high-risk markets in the wake of the financial crisis in Asia.[36] Given the inefficacy of tax collection and the withholding of a loan tranche worth US$700 million by the International Monetary Fund (IMF) because tax collection targets had not been met, privatization receipts appeared to be the only way to settle the politically sensitive issue of US$1.6 billion in salary arrears to state employees, which Yel'tsin had promised would be paid by early 1998. In fact, in order to attract as many bids as possible (and hence increase revenues received from sales) for stakes in Slavneft' and Rosneft', Yel'tsin issued a decree which removed the previous 15% limit on foreign ownership of oil companies.[37] While it is unfair to blame the government for not foreseeing the deterioration in market conditions, it is possible that the state's shares in Slavneft' might have been sold if a larger stake had been on offer.

The third factor that influenced the outcome of the Slavneft' sale in 1997–1998 was the attitude of the company's management towards privatization. Fomin was supportive of privatization in principle, but only if his preferred option was implemented. This involved allowing a group of foreign institutional investors – comprising large European banks with whom negotiations were already well under way – to acquire the stake in question. According to Romanov, this plan was ideal because it allowed Fomin to 'privatise the company with somebody else's money but according to management's interests, whereby the shareholder would not interfere with the business process'.[38] In return for their hands-off approach, foreign investors were guaranteed a 40% annual return on the value of the company's shares, while management would retain 20% of capital gains. For TNK, the appearance of this alternative plan made it all the more determined to abrogate the auctions since it could not hope to compete with the terms offered to Fomin by the foreign investors.

From the above account, it is clear that all three elements of the triangular relationship – the tycoons, bureaucratic agencies and management of Slavneft' – interacted in such a way as to minimize the chances of a successful divestiture of shares. Although the government was keen to sell the shares, opposition to the sale by some tycoons and by Fomin's support for the alternative plan meant that the attempts to privatize Slavneft' in 1997–1998 were doomed to fail. Claims by analysts and even senior privatization officials in Russia that the Slavneft' auctions were affected by a 'depression on Russia's financial markets brought on by the Asian markets crisis'[39] were simply inaccurate, since the downturn in the world financial markets was merely a catalyst.

1999–2001: Covert posturing in anticipation of privatization

Between 1999 and 2001, the government approved annual privatization programmes that envisaged the sale of 19.68% of Slavneft'. However, no formal announcement of such a sale was actually made. The retention of the status quo at Slavneft' during this period was not primarily due to poor investor sentiment as a result of Russia's default and ruble devaluation in August 1998. Neither was it directly linked to political uncertainty in the country, including the surprise resignation of Yel'tsin in 1999, nor the nomination of a relative unknown as acting prime minister and fears that the renewed war against Chechnya would result in attacks on people living in Moscow. In fact, despite poor market conditions and political uncertainty during those years, GKI continued to divest successfully its shareholdings in a range of oil, coal and insurance companies. It is also not quite accurate to ascribe the government's decision to postpone the privatization of Slavneft' to internal divisions about the merits of merging the remaining state-owned oil companies, including Slavneft', into a single national oil company or Gosneft'. While these debates were often fierce and inconclusive, they did not prove to be a significant obstacle to the privatization of ONAKO, which was one of the companies that were supposed to comprise Gosneft'. In short, any direct link between the government's decision to postpone privatization in the face of uncertainties – such as lacklustre demand, changes in the political climate and debates over the formation of

Gosneft' – and progress in privatization is, at best, a tenuous one. Rather, as will be explained later, the loss of momentum in the privatization of Slavneft' may be directly attributed to the role played by the company's management.

As for private business groups, they were badly affected by the economic downturn. For example, the Russian stock market index plunged 90% in a space of one year between October 1997 and September 1998.[40] Given that oil and gas companies accounted for 55–65% of the capitalization of the stock market, its dramatic fall severely affected the value and purchasing power of companies that were potential buyers of Slavneft'.[41] As such, the leaders of oil companies were more intent on saving and consolidating their existing business empires rather than acquiring new assets: this partly explains why they declined the opportunity to purchase Rosneft' in 1998, and why they did not pressure the government to implement its privatization plans for Slavneft'.

However, the financial situation at oil companies began to improve in 2000 partly as a result of higher oil prices, which rose steadily from a low of US$12 per barrel in 1998 to US$23 per barrel in 2001.[42] They also found it easier to raise funds from Western investment banks as a result of adopting moves towards greater transparency and corporate governance, for instance, by implementing internationally accepted accounting standards. Any doubts about the attitude of the new president towards privatization also eased after mid-2000 following a meeting with Putin, during which the tycoons were told that there would be no revision of past privatization results if they desisted from politics. In a clear indication that by 2000, private business groups had recovered from the financial meltdown, the leading oil companies embarked on a buying spree. At the end of 1999, LUKoil acquired regional oil company, KomiTEK, in a share swap worth US$500 million. YUKOS increased its stake in oil company VNK from 20% to 68% during the last quarter of 2000. As for the Al'fa group, it purchased 85% of a minor oil company, ONAKO, for US$1.08 billion in September. This was in addition to Al'fa's purchase of two key subsidiaries of a bankrupted oil company, SIDANKO, in late 1999. Al'fa was therefore hardly short of financial resources. In this regard, the reason why it failed to persuade the government to put Slavneft' up for sale, despite its obvious interest in acquiring the company, had more to do instead with management intransigence at Slavneft'.

The management of Slavneft' continued to object to privatization and, indeed, to any change in the company's status as a state-owned enterprise. Vasiliy Duma, who replaced Fomin as president of Slavneft' in August 1998,[43] refused to confirm the valuation of 19.68% of the company's shares carried out by foreign assessors at the request of RFFI.[44] He also spoke out against plans to include Slavneft' in Gosneft' and was removed by Mintopenergo, which was 'displeased with the increasing independence of the company'.[45]

His successor, Mikhail Gutseriev, likewise frustrated efforts by the government and tycoons to privatize the company during the first two years of his tenure. For example, shortly after his appointment in January 2000, he argued that it was 'senseless' to form a state oil company when 90% of Russia's oil sector was already privatized, opined that the possession of a 19.68% stake was 'meaningless' since the government should sell 'either everything or nothing' and added that he would not agree to any privatization plans until after 2001 at the earliest since management should be given time to implement plans to increase the capitalization of Slavneft'.[46]

He also rebuffed attempts by Al'fa group's TNK to create a more favourable environment within Slavneft' for privatization by placing its own representatives on the board of directors of Slavneft' and its subsidiaries; indeed, by mid-2000, TNK and its partners had acquired 12.8% of Slavneft' along with 30% of its key oil producer (Megionneftegaz) and refinery (Yaroslavnefteorgsintez).[47] However, Gutseriev's team 'did everything possible and used every existing loophole in Russian legislation to debar minority shareholders from grasping any levers of management'.[48] For instance, as a result of criminal investigations by the Interior Ministry into illegal sale of shares by Mozyr, the stakes in Slavneft' held by TNK and its partners were frozen ahead of a crucial annual shareholder's meeting to decide on the composition of the board. This resulted in TNK and its partners being barred from participating in the meeting. Although a municipal court later ruled that the seizure was invalid, the ruling came too late for TNK to participate in the meeting.

It appears that Gutseriev managed to use his close relations with the Interior Ministry to deploy the coercive capacity of the state for his own private benefit, as the above example demonstrates. He also allegedly instigated a police raid of TNK's branch offices[49] as part of a harassment campaign against the company. Indeed, it has been claimed that Gutseriev's nomination as president of Slavneft' had been supported by its minister Vladimir Rushailo in return for a portion of the company's profits.[50] As we will see, Gutseriev repeatedly relied on organs of the state as a *krysha* or private protection arrangement, in his struggle to retain his position at Slavneft'.[51]

Apart from frustrating the attempts of tycoons to engage in 'creeping privatisation', through changes to the composition of the board of directors, Gutseriev also managed to persuade GKI not to proceed with divesting the 19.68% stake in 2000, on the basis that the new management team should be given a chance to carry out its projects aimed at increasing the capitalization of the company.[52] The government would then be able to sell its stake for a higher price at a later stage. The announcement by GKI in July 2000 that Slavneft' would not be put up for sale in 2000 was greeted with disappointment by TNK, since the stake would have increased its control over Slavneft'.

Gutseriev even managed to persuade the Duma to enact legislation banning the sale of Slavneft' in 2001. In September 2000, GKI submitted a draft law on privatization aimed at clarifying decision-making authority and criteria for appraising state property, in a bid to minimize the annual confrontation with the Duma over enterprises to be targeted for privatization. Since the proposed law would deny the Duma the right to approve lists of strategic enterprises for privatization and thereby reduce its control over privatization,[53] the Duma responded by enacting legislation that prohibited the sale, in 2001, of state-owned shares in enterprises, whose assets were more than 50 million times the minimum wage, pending the adoption of the new federal law on privatization. As a consequence, Slavneft' was excluded from privatization in that year. The new law on privatization was only adopted in December 2001, thereby opening the way for the privatization of Slavneft' and other large enterprises the following year.

Gutseriev's complicity in this affair was based on the fact that prior to his appointment at Slavneft', he had served as deputy speaker of the Duma and was probably able to make use of his 'close ties ... with State Duma deputies, his former colleagues'.[54] Moreover, Gutseriev is independently wealthy – he founded the BIN conglomerate in 1992, which has interests in oil, banking and real estate, and is one of the 100 richest people in Russia – and it is not inconceivable that money was exchanged for votes in the Duma. In fact, he had bought himself a top spot on the Liberal Democratic Party's party list by financing its successful electoral campaign of 1995.[55] Gutseriev enjoyed the support of the Duma on other occasions as well. For example, in the midst of Gutseriev's altercations with TNK over representation on the board of Slavneft', the Duma instructed the Audit Chamber to investigate the validity of a transaction in which the Al'fa group had acquired 50% of TNK's shares from the state. In addition, in mid-2002 when Gutseriev's position was being challenged by Yuriy Sukhanov, the vice-president of

Slavneft' and a former employee of Sibneft', the Audit Chamber and the Interior Ministry launched an investigation into large-scale embezzlement by Sukhanov, who had supposedly used his position in Slavneft' to sell oil cheaply to a Sibneft' affiliate, which then resold it at higher, world market prices.[56]

The 'principal–agent problem' provides a partial explanation for the behaviour of Gutseriev and his predecessors at Slavneft'. The problem refers to a situation which arises when the objectives and interests of a manager to whom a task is delegated (the 'agent') are not aligned with those of his employer (the 'principal') whose interests he is supposed to serve. This is because being a rational, self-interested individual, the manager would seek to maximize his own preferences with minimal costs to himself, that is, by using resources owned by his principal but under his control. In an attempt to minimize such discrepancies, the principal is obliged to incur 'agency costs' including the use of moral, material and coercive incentives to motivate the agent, gaining information to set performance targets and implementing performance-monitoring techniques. In this connection, well-designed institutions (formal rules and informal norms) can provide the correct incentives to ensure that decisions taken by the agent would be in harmony with the interests and expectations of the principal.[57]

Gutseriev's personal, pecuniary interests were out of step with those of his principal, the state, whose resources he was using for his own ends. As president of Slavneft', he controlled the company's cash flows, voting rights and agendas at shareholder's meetings. As mentioned earlier, this allowed him to divert part of the profits of Slavneft' to himself and his backers, namely, Rushailo, an oil trading company linked to Sibneft', and a criminal group. It also benefitted Gutseriev's own BIN conglomerate since the accounts of Slavneft' were transferred to BIN bank, which subsequently became the company's landlord.[58] Later in 2002, Gutseriev became the co-owner of an investment trust that held 13% of the shares of Slavneft' – owned in equal parts by TNK, Sibneft' and two companies affiliated with Gutseriev[59]. Subsequently, Gutseriev managed to gain ownership of Varyegannneft', an oil-producing subsidiary of Slavneft' which accounted for 10% of the holding company's total production and 12% of its reserves.[60] It became the core of Russneft', a wildly successful but short-lived VIOC created by Gutseriev following his ouster from Slavneft', which he later sold for a huge profit.[61]

Unfortunately for the state, the few institutions that were available to minimize the principal–agent problem were largely ineffective. Material incentives provided by the state paled in comparison to what Gutseriev managed to generate for himself on the sly, while coercive incentives in the form of dismissal did not concern him since, in his own words, 'I receive proposals all the time from different organizations, inviting me to work as a leader of the very highest rank.'[62] As for the state representatives on the board of directors, they had few legitimate grounds to seek his dismissal since the board's chairman, who was the head of GKI, declared himself to be satisfied with the 'excellent' results of the company's activities.[63] Moreover, Gutseriev received the support of the Belarussian representative on the board since Belarus was strongly opposed to the privatization of Slavneft'.

The other reason why Gutseriev was able to 'defect' from cooperating with his principal on privatization concerned the political nature of Gutseriev's appointment. Specifically, Gutseriev was appointed to head Slavneft' in order to ensure the loyalty and allegiance of one of the leading clans in Ingushetia, prior to the Kremlin's intensification of military operations in neighbouring Chechnya in 1999. This clan comprised the Gutseriev brothers and the president of the republic, Ruslan Aushev. As previously mentioned, Rushailo backed the nomination of Gutseriev at Slavneft', and it cannot be mere coincidence that Rushailo's Interior Ministry was also heavily involved in operations in Chechnya.[64] Gutseriev, as deputy speaker of the Duma, was the highest-ranking Ingush within the Russian state apparatus and was widely regarded as the 'co-owner of Ingushetia'[65] in view of the prominence of his BIN group in the republic. The Russian government under Yel'tsin had, in fact, used a similar tactic to buy off the Ingush (who are ethnic kin of the Chechens) prior to the invasion of Chechnya in 1994, when it decreed the creation of a free economic zone in Ingushetia administered by BIN bank and, thereafter, transferred huge amounts of federal subsidies to the republic over a number of years.[66] Coupled with the fact that all previous presidents of Slavneft' were senior bureaucrats from Mintopenergo, it is obvious that Gutseriev's appointment was politically motivated and that allowing him to head and remain at Slavneft' was deemed to be vital since it minimized the chances that the Russian army would have to fight a two-front war in Chechnya and Ingushetia. Gutseriev himself admitted that his appointment was more a 'political post',[67] a claim later repeated by one of his closest aides who remarked that Slavneft' was 'more about politics than economics'.[68] In this respect, Gutseriev's personal agenda and the widening agency problem at Slavneft' was tolerated for the time being.

Privatization in 2002

A solution to the above principal–agent problem soon appeared in the guise of presidential elections in Ingushetia which were called when Aushev announced in December 2001 that he was stepping down prior to the end of his term. Gutseriev involved himself in the electoral campaign to ensure that his brother, Khamzat, the republic's Minister of Internal Affairs at that time, was registered as a candidate. This was despite warnings that he was putting at risk his own position at Slavneft'.[69] As for Aushev, he openly campaigned on behalf of Khamzat Gutseriev. The Kremlin fielded its own nominee, Murat Zyazikov, who won the elections after Khamzat Gutseriev's candidacy was declared illegal by the courts on the grounds of bribery and procedural violations a few days before the final round of voting in April 2002. Subsequently, at a shareholder's meeting in mid-May, Gutseriev was dismissed as president of Slavneft' amid allegations that he had used the company's resources to finance his brother's campaign.[70]

For Putin, who was two years into his first term, the dominance of the Gutseriev–Aushev clan was becoming a liability. They not only ruled Ingushetia as a fiefdom, but were also suspected of perpetuating the war in Chechnya by sheltering and financing Chechen rebels and also by opposing the return of Chechen refugees from Ingushetia in order to continue receiving (and pocketing) federal aid to the refugees.[71] The fact that Chechnya still refused to yield to Moscow's authority was especially galling because the war there had been the 'opening gambit in a systematic effort to restore state authority and the 'power vertical' in the Russian political system'.[72] Indeed, the failure in Chechnya contrasted sharply with success in other spheres, where Putin managed to appoint presidential plenipotentiaries to super-regions of Russia, dismiss governors from their posts, reduce the power of the Federation Council as a seat of gubernatorial influence, abrogate or revise bilateral federal–regional treaties concluded under Yel'tsin's administration, dissuade some business tycoons from pursuing overt political aims and assert state control over much of the media.[73] In order to normalize the situation in Chechnya ahead of parliamentary and presidential elections in 2004, talks between Russian and Chechen representatives took place in October 2001 for the first time since the start of the war. Aushev was forced to resign, and Zyazikov was elected as the new president. The ouster of Khamzat Gutseriev, from the presidential elections, and his brother, from Slavneft', was part of this tentative but short-lived attempt to eliminate obstacles to a political solution in Chechnya. Moreover, given that the Duma's ban on the privatization of Slavneft' was no longer applicable in view of the new privatization law signed by Putin in March 2002, the costs to the government of keeping Gutseriev at Slavneft' were no longer acceptable.

In May 2002, Gutseriev was replaced by Sukhanov, the aforementioned vice-president of Slavneft' closely allied with Sibneft', who promptly brought over four senior managers from Sibneft' to his new team at Slavneft' and declared his intention to prepare the company for privatization scheduled later in the year. According to the privatization programme approved by the government in 2001, Slavneft' would be privatized in two stages: a 19.68% stake would be sold in the autumn of 2002, while a second stake of 5.2% minus one share would be divested in 2003. The remaining 50% plus one share would be retained by GKI. This plan was reiterated by the government in July 2002, and in August, consultants hired by the RFFI provided a valuation of the company based on the above plan. However, in October 2002, the government suddenly announced a new privatization plan for the company, whereby the entire 74.95% stake held by the Russian government would be sold as a single lot by the end of the year. How and why did this sudden and unexpected change occur?

According to a leading Russian journalist, 'it was Abramovich who pushed for the Slavneft' auction'.[74] This observation is only partly true: it overlooks the important role played by TNK and its parent company, the Al'fa group, acting in concert with Abramovich to lobby the government for a change in the privatization programme. It is worth recalling that TNK and Sibneft' had, in fact, been bitter rivals and had fought a few corporate battles over acquisitions in the oil industry. As mentioned earlier, TNK's attempts to seek representation on the boards of Slavneft' and its subsidiaries were rebuffed by Gutseriev, who had by then established a mutually profitable relationship with Sibneft'. The second area of contention between TNK and Sibneft' was over ONAKO, 85% of which was bought by the Al'fa group in September 2000 from the state. However, at around the same time, Sibneft' managed to acquire a 42% stake in ONAKO's only oil-producing subsidiary, Orenburgneft, thereby complicating TNK's control over ONAKO.

During 2002, however, there was a gradual rapprochement between TNK and Sibneft', which was shaped by the widely anticipated privatization of Slavneft'. In late April 2002, TNK sold its stakes in Slavneft' to an investment trust owned by Sibneft' and Gutseriev. As a result, the trust controlled 13% of Slavneft' in addition to blocking stakes in other subsidiaries, and its creation was perceived to be a 'joint defense of Slavneft' and its subsidiaries' in order to 'fend off other bidders'.[75]

The threat from rival bidders was a very real one, as exemplified by the case of Sergey Pugachev's Mezhprombank. Between April and July 2002, Mezhprombank was involved in a protracted management struggle at Slavneft' when it supported Gutseriev's attempts to retain his position.[76] This brought Mezhprombank into direct conflict with Sibneft', which had thrown its weight behind its own man Sukhanov. Prior to the shareholder's meeting in May that officially terminated his appointment as president, Gutseriev took a leave of absence and appointed Anatoliy Baranovskiy, a board member with close ties to Mezhprombank and a vice-president of Rosneft', as acting president of Slavneft'. By then, two other Mezhprombank employees had already joined Slavneft' as vice-presidents.

Following Sukhanov's election as president in May 2002, Baranovskiy and bodyguards from BIN-Zashchita, a security firm owned by Gutseriev, occupied the offices of Slavneft'. A writ was produced according to which a court in U'fa, the capital of Bashkortostan and where Pugachev's partner was running for president, annulled the recently concluded shareholder's meeting.[77] A month later, Gutseriev himself entered the building while troops from the Interior Ministry, BIN-Zashchita and Interbezopasnost' (a private security firm owned by Pugachev) prevented his rival, Sukhanov, from entering the office. After three days, Sukhanov took control of the premises with the help of security personnel from a firm owned by Sibneft'. At that time, Pugachev was widely assumed to be part of the *siloviki* clan,[78] representatives of which were trading on their close ties to Putin to make inroads into the top echelons of the bureaucracy and presidential administration and who therefore posed a challenge to the dominance of people (dubbed 'The Family') who had benefited politically and materially under Yel'tsin's administration. Members of 'The Family' included Abramovich, the then prime minister Mikhail Kas'yanov and Aleksandr Voloshin, the head of the presidential administration at that time. In this regard, Sibneft' calculated that it was in its interests to work with TNK to lobby for a change in the privatization programme. This was because under the existing plan, Mezhprombank would be able to finance the bids for Slavneft' since the stakes and prices in question were relatively small. However, Mezhprombank would be financially disadvantaged if a larger and more expensive stake, which the joint resources of Sibneft' and TNK could afford, was for put up for sale instead.

Perhaps even more worrisome for TNK and especially Sibneft' was the possibility of a combined bid for Slavneft' by Mezhprombank and Rosneft', which appeared to have formed a strategic alliance. In February 2002, for instance, Rosneft' purchased a 50% stake in Mezhprombank's Russkiy Ugol', which then proceeded to expand its portfolio of coal companies in Russia and Ukraine. Also, during the above-mentioned management dispute at Slavneft', there was some suspicion that the president of Rosneft', Sergey Bogdanchikov, had in fact supported the candidacy of Baranovskiy as a step towards eventually including Slavneft' into a national oil company led by Rosneft'.[79] It therefore appeared that Bogdanchikov had become 'an integral part of the so-called BMP group of Messrs Bogdanchikov, Miller [Aleksey Miller, head of Gazprom] and Pugachev',[80] which made it all the more prudent for TNK and Sibneft' to cooperate on matters related to Slavneft'.

In October 2002, TNK and Sibneft' were brought closer together following the conclusion of a settlement over ONAKO.[81] This settlement was actually concluded a week before it was publicly announced on 15 October, at around the same time as GKI's submission of a draft privatization plan for Slavneft' to Prime Minister Mikhail Kas'yanov for consideration. Bearing in mind this time frame, it is likely that Abramovich's relationship with Kas'yanov as members of 'The Family' accounted for the 'lightning speed' at which the proposal was endorsed.[82] At the same time, Voloshin, who was 'The Family's' representative in the presidential administration, and Al'fa's own links to the presidential administration (Vladislav Surkov and Aleksandr Abramov, deputy chiefs of Putin's administration, were previously senior managers at Al'fa bank)[83] helped to elicit a rapid approval of the new plan from an authority 'higher than just GKI, namely the Kremlin: it came from there'.[84] The rapprochement between TNK and Sibneft', which arose because of the development of their business relationship and the new threat posed by rival bidders, must therefore be seen as the key factor that led to the adoption of a new privatization plan for Slavneft'.

Thus far, the analysis has focused on the motivations and behaviour of tycoon-owned groups and the new managers of Slavneft' to agree to the privatization of Slavneft' in 2002. It is now timely to explore the role of state officials in this respect. In a reference to the drama surrounding the management dispute at Slavneft', GKI explained that the new privatization plan for the company was adopted because 'the example of Slavneft' showed that the state [was] not an effective manager. Two presidents appeared and court decisions were implemented with the help of law-enforcement bodies'.[85] However, the lament about the state as manager is not a new one: since 1992, it has been cited as the key reason to reduce state control over enterprises.[86] In addition, empirical surveys have reinforced this justification and demonstrated that in general, privately owned firms outperform state-owned enterprises, even in Russia.[87]

Furthermore, the use of courts and security forces in business disputes has hardly been limited to Slavneft'. Yakov Goldovskiy – the former chief executive of Sibur, which is the main producer of petrochemicals and is 51% owned by Gazprom – managed to acquire for himself subsidiaries of Gazprom with the help of US$120 million of the latter's own money. Likewise, as noted in Chapter 1, former managers of Rosneft' managed to buy the company's key oil-producing subsidiary, Purneftegaz, for US$10 million in a forced sale resulting from overdue debts of US$6 million, even though it was worth at least US$500 million. Yet, unlike Slavneft', the state did not react to these instances of managerial inefficiency by privatizing Gazprom or Rosneft'. The explanation by GKI therefore does not account for the timing of the introduction of the revised plan for Slavneft', nor the choice of Slavneft' as a target for privatization as a way to improve professionalism.

The other reason offered by the state to explain the new terms for the sale of Slavneft' concerned Russia's foreign debt payments due in 2003. Apparently, the government decided to dispose of the entire oil company, instead of a 20% stake as was originally planned, because this would 'compensate for the failure of the major privatization project of the year: distribution of … 5.9% of LUKoil shares at the London stock exchange' in August 2002.[88] Indeed, Kas'yanov ordered that the proceeds from the sale were to be received by the government by 15 February 2003, exactly the same day that Russia was due to make the US$1.5 billion repayment to the IMF.[89] However, the issue of debt repayment was probably only a convenient justification to sell Slavneft'. A former deputy head of GKI revealed that 'if Slavneft' had not been sold, the debt payments would still have been made in time. We should not imagine that shares were sold only to receive money. When Slavneft' was put up for sale, there was no talk of selling it because we needed to pay our debts. What was said was that it was necessary to sell it and as a consequence, this would make it easier to pay the debts'.[90]

Why then, was it really necessary for the state to sell Slavneft'? The answer lies in the self-regarding interests of its top political figures, Kas'yanov and Putin, and the constraints placed on them by the existing rules of the political game, according to which power lay with 'The Family' at that point in time. Given that Kas'yanov aspired to the presidency, he calculated that he could improve his standing within 'The Family' by 'making every effort to ensure that Slavneft' [would] go to the Family's chief financier, Roman Abramovich ... [so that] as much as possible of Russia's resources [would be] concentrated under the Family's control'.[91] In this way, as noted by a Russian newspaper, 'should Putin prove recalcitrant, then his rival in the presidential race may well turn out to be Mikhail Kas'yanov'.[92]

In Putin's case, he was no doubt painfully aware that he lacked a significant power base of his own and was, to a large extent, beholden to 'The Family' for much of his first term. Nevertheless by not objecting to Kas'yanov's sudden turnaround to privatize all of the state's shares in Slavneft' and by not standing in the way of Abramovich's desire to acquire the company, Putin bought himself the continued support of 'The Family' for his tenure as president and won for himself more time to appoint his own coterie of supporters to key positions, which eventually allowed him to become more independent of 'The Family'. For Putin, therefore, his agreement to the Slavneft' sale was a strategy to simultaneously manage and subsequently create a new coalition of supporters.

The auction for the Russian government's 74.95% stake in Slavneft' took place on 18 December 2002. On the one hand, it was more professionally organized than hitherto, since unlike the earlier YUKOS auction in December 1995, bidding was conducted under the direct auspices of RFFI and not contracted out to an organizer linked to a participant. On the other hand, the auction was criticized for the fact that 'one giant bidder after another dropped out of the game' so that 'Slavneft' was divided in a matter of minutes between Sibneft' and TNK'.[93] Nevertheless the diminishing pool of bidders in the run-up to the auction was, for the most part, a rational reaction to the successful bid by Sibneft' for the 10.85% stake of Slavneft' owned by the Belarussian government. This discouraged potential bidders from participating in the Russian auction, because it meant that Sibneft' now had close to a blocking stake in Slavneft' – that is, 10.85% plus the 13% share in the investment trust – and could demand a prohibitively high price in the event of a buyout by the majority owner. The winner of the Belarussian auction was announced on 8 December, but rumours had begun circulating before then. The fact that LUKoil and Surgutneftegaz withdrew from participation, just around the time of the official announcement by Belarus, suggests that this victory by Sibneft', rather than covert manipulation, was the key determinant of their decision.

As a result, Sibneft'–TNK easily emerged as winners at the auction since the other participants were companies affiliated to them. Thus ended the torturous journey for Slavneft' from a state-owned company to a privately held enterprise owned in equal parts by Sibneft' and TNK (the latter was renamed TNK-BP following a successful buyout in 2004).

Observations on decision-making

YUKOS versus Slavneft'

The Menatep group was not always the driving force behind the privatization of YUKOS particularly in the earlier phase, where the motivations of the state and internal management were just as significant. More so than in the case of YUKOS, the behaviour of the management of Slavneft' was in fact a key determinant of the extent to which the corporatization and privatization of Slavneft' was successful. Fomin, for example, was the driving force behind the creation of the company, while Gutseriev's repeated attempts to frustrate potential buyers accounted in large part for the delays in privatization. Their behaviour was not the result of an intrinsic belief in a large role for the state in the economy, but was grounded in the pursuit of private, self-regarding preferences for material gains. The tendency to focus only on the machinations of tycoons must hence be reconsidered.

Another point of comparison between the privatization of YUKOS and Slavneft' concerns the role of officials from Putin's presidential administration and their allies in state-owned companies. Specifically, why did they manage to obtain control over Yuganskneftegaz in 2004 but fail to do likewise with Slavneft' two years earlier? The reason for this relates to Putin's coalition-building strategy which is aimed at prolonging his political survival. As explained previously, the lack of his own power base for much of his first term obliged Putin to be responsive to the interests of 'The Family', which included the acquisition of Slavneft' by Sibneft', in order to ensure his political longevity. Subsequently, having hand-picked his own coalition of supporters by the second term, Putin allowed them to wrest control of Yuganskneftegaz to consolidate the alliance and to tie them even more to his own political future.

Institutions and actors

None of the actors involved in the case study of Slavneft' – the state, big business or internal management – dominated the triangular relationship all the time. The reason for fluctuation stems from the different possibilities of relationship between institutions, defined as formal rules and behavioural norms, and individual decision-makers in a given contextual setting. In the first place, individuals can work within exist-ing institutions to achieve private aims. In Fomin's case, he lobbied for the creation of Slavneft' as a project in line with Russia's new foreign policy of closer relations with the CIS countries and, in particular, with Slavic states such as Belarus. Secondly, institutions can affect decision-making by individuals. For instance, the poor incentive structure and weak enforcement capabilities of the state in the principal–agent rela-tionship, encouraged Gutseriev to act as on obstacle to its plans to pri-vatize the company. Third, actors can attempt to change institutions in order to affect the behaviour of others. TNK and Sibneft', for example, decided to create a new alliance and to use their combined influence to lobby for the adoption of a new privatization plan for Slavneft'. By changing the existing rules of the game, TNK and Sibneft' successfully discouraged their rivals from making a bid for Slavneft'.

In fact, the different ways in which institutions interact with actors underline the insightful argument by Herbert Simon that explaining individual behaviour within organizations solely in terms of agency, transaction costs or incentives ignores the significance of 'soft' factors such as organizational loyalty.[94] As managers of state-owned oil com-panies, Muravlenko of YUKOS and Fomin and Gutseriev of Slavneft' operated within similar institutional parameters. However, whereas Muravlenko facilitated the attempts by the state to privatize the com-pany, Fomin and Gutseriev hindered such attempts. In all likelihood, the difference in outcome had to do with the extent of organizational loyalty. Muravlenko was part of the old, paternalistic school of Soviet managers who prided themselves on improving the lives of their employees through social amenities, and who was aware that he was unable to lead the firm under market-driven conditions. Fomin and Gutseriev, on the other hand, were more concerned with profitability and efficiency, and how they could use their appointments for private gain. Indeed, senior executives in the oil industry consistently spoke about Fomin with awe and in hushed tones, whereas they regarded Muravlenko as an affable person. Muravlenko was therefore more pre-pared to work with the state and Menatep to ensure the financial and operational viability of YUKOS.

External influences

Finally, the case study of Slavneft' has contributed to an appreciation of how external and global influences impact on privatization in Russia. We saw earlier how the global popularity of privatization programmes did not fundamentally determine the choice by the 'young reformers' to introduce market reforms in Russia. Rather, it was their experience with the failures of a command economy and of previous attempts at reform communism that was the driving force: the worldwide trend in favour of privatization merely confirmed their prior inclinations. The analysis in this chapter – namely, how the debt payment to IMF was used by the state to justify the new privatization programme for Slavneft', and how the weakness of world financial markets was used to explain the failure to privatize the company in 1997 and 1998 – reinforces the point about the limited role of external influences on decision-making with regard to privatization. In Russia, domestic actors and their self-interests remain the key determinants and drivers of privatization, rather than institutions or externally generated influences.

5
Case Study of Rosneft'

The previous case studies identified the primary role played by the incumbent general managers during the initial phase of privatization. The examples of YUKOS and Slavneft' also underlined the contention that the eventual sale of these companies was only possible when the three key actors – the general manager, the state and big business – concurred on the issue at the same moment in time. This analysis of Rosneft' will examine the extent to which these conclusions continue to be valid in the case of an oil company that was created in 1993, corporatized in 1995 and put up for sale three times in 1998 but without success. In connection with the partial sale of Rosneft' in July 2006, this chapter will also consider why privatization succeeded this time around.

Creation and corporatization

The creation of Rosneft' was, in many ways, a by-product of the consensus that was gradually forged among key representatives of state agencies, the oil industry and the non-oil sectors regarding the necessity of creating a holding company in the early 1990s. Such a company could immediately absorb and claim ownership over that part of the oil industry that remained outside vertically integrated structures and hence reduce the incidences of 'spontaneous privatization'. The consensus, however, stopped there. For some key actors, any such holding company would merely be a stopgap measure. GKI, for example, perceived the holding company to be a temporary structure, whose assets could be further subdivided to form new VIOCs; in this way, competition and efficiency would increase, and state control over the economy would decline. Similarly, ambitious 'oil generals' regarded the holding company as a transient structure to be later subdivided into independent entities for their own self-enrichment. For those opposed to de-monopolization and general privatization, it was envisaged as the basis of a national oil company, which would guarantee a cheap supply of energy and exist alongside a handful of corporatized VIOCs. In any case, the debate focused on whether this oil holding company should be created from scratch, or whether the existing Rosneftegaz industry association could be reorganized to play such a role.

Rosneftegaz, as mentioned earlier, was a voluntary association comprising mainly oil production and refining companies that had been created in October 1991 from the Soviet Oil Ministry. Its reorganization at that time had been motivated by a rational sense of bureaucratic self-preservation. For instance, senior officials of the Oil Ministry worried that 'if we did not take the initiative, Yel'tsin's government would put together its own ministry. What would become of us then?'[1] This same sense of self-preservation lay behind Rosneftegaz's attempt to position itself, through a report published by consultants it hired, as the proposed holding company for the oil industry.[2]

However, as a result of decree 1403 issued in November 1992, a new company was formed to take on the role of the proposed holding company. The reasons for this outcome lay partly in the fact that Rosneftegaz suffered from an 'image problem'. Specifically, it was perceived as merely the much-criticized Soviet Oil Ministry under a new name and was thus written off as hopelessly inefficient and resistant to change: the head of Rosneftegaz, Lev Churilov, was in fact the last Soviet Oil Minister.[3] Also, a significant number of Rosneftegaz's staff had previously served at the Ministry. Another reason for Rosneftegaz's demise concerned the fact that it was engaged in, and eventually lost, a 'guerilla warfare between its management and the [Fuel and] Energy Ministry over the definition of powers and responsibilities'.[4] As noted in Chapter 2, Mintopenergo faced numerous calls for its own dissolution throughout 1992, in view of its seeming inability to reverse the steep decline in oil production and reduce corruption with regard to its jurisdiction over tax-free export quotas. Mintopenergo therefore considered the existence and ambition of Rosneftegaz as a threat to its own institutional viability. It was hardly surprising that upon his appointment as head of Mintopenergo in January 1993, Shafranik consolidated his position with firstly, the removal of Churilov as head of Rosneftegaz, secondly, the dissolution of Rosneftegaz and finally, the appointment of Aleksandr Putilov (who was Shafranik's former colleague at LUKoil when they were general managers of the company's main production units) as president of the newly created oil holding company, Rosneft'.[5]

Rosneft' received into trust management the state's shares in 259 oil-related enterprises that were not part of the three VIOCs (LUKoil, YUKOS and Surgutneftegaz) created by decree 1403. It was tasked to restructure and distribute the 259 enterprises into new VIOCs, coordinate state investment in the oil industry, ensure a steady supply of oil and oil products for state needs and facilitate international trade.[6]

Subsequently, with the proliferation of new VIOCs between 1993 and 1995, the number of enterprises under the direct control of Rosneft' decreased significantly and sparked a debate on the future of the company. Putilov opposed the formation of ever greater numbers of VIOCs and suggested that there should only be 'seven to nine major integrated companies . . . and one Russian national company, Rosneft''.[7] He also fought to retain the company's best assets. For instance, in mid-1994, he complied only reluctantly and belatedly with GKI's order for Rosneft' to transfer the state's shares in an oil-producing enterprise, Purneftegaz, to a newly created VIOC, SIDANKO.[8] He then persuaded the Prime Minister Chernomyrdin to issue a new resolution that nullified GKI's order and returned control of Purneftegaz back to Rosneft' in January 1995. Given that a decision had already been made by then to allocate a large oil-production unit, Nizhnevartovsk, controlled by Rosneft' to a new VIOC, the loss of Purneftegaz would have relegated Rosneft' from an oil company able to compete with the large VIOCs to a second-tier enterprise. As will be evident, this need to justify and fight for the mere existence of Rosneft' as a company has been a recurring theme throughout much of the company's history.

Groups that opposed the designation of Rosneft' as a national oil company felt that there was no need for such an entity. LUKoil, YUKOS and SIDANKO argued that such a company would have unfair advantages in terms of its preferential relationship with the state.[9] Mintopenergo also questioned the need for a national oil company, since the latter's wide-ranging powers over the industry would undermine the raison d'être of the Ministry. In this connection, it declared that if, by chance, a national oil company was actually established, the latter would merely 'carry out purely co-ordination work'.[10] Therefore, for opponents of Rosneft', the company could either be transformed into just another corporatized VIOC as the first step towards privatization, or be dissolved once all its productive assets had been allocated to existing VIOCs. Indeed, having reviewed the situation in the oil industry as at mid-1994, a senior GKI official concluded that since 'the presidential decree [of 1403 was] 94% fulfilled, the end of Rosneft' [was] not far away'.[11]

The outcome was a political fudge. Presidential decree 327, dated 1 April 1995, 'On Top Priority Measures to Improve the Performance of Oil Companies' transformed Rosneft' into a corporatized entity. According to the proposed privatization plan submitted with the charter, the structure of Rosneft' would be similar to that of other VIOCs, where 51% of the shares of the holding company would be retained by the state for three years. In addition, 25% of the company's shares would be distributed at a discounted price to its employees and 24% would be sold to outside investors.[12] It was not designated a national oil company, but was instead given some of the privileges and responsibilities commonly associated with such a status. For example, Rosneft' was tasked with managing state-owned shares of oil enterprises that were not part of any other existing companies and with coordinating scientific research for the oil industry. It also received the exclusive right to negotiate and manage the state's share of oil and gas from projects carried out under the auspices of production-sharing agreements with foreign partners.

This compromise was a reflection of political realities at a time when the authority of the government was too weak for it to give outright support to any one group in the above debate. In the wake of the financial crisis in October 1994 – when the ruble dramatically lost a quarter of its value against the dollar in a single day – senior officials were sacked, Chernomyrdin himself was nearly forced to resign, and the government narrowly survived a no-confidence vote in the Duma. The compromise over Rosneft' thus allowed Yel'tsin and Chernomyrdin to be seen as embracing as many constituents as possible, ahead of parliamentary elections in December 1995. As a result, Rosneft' was once again saved from obsolescence.

Thus far, the analysis of the origin and corporatization of Rosneft' suggests that unlike YUKOS and Slavneft', the company's own managers were less able to influence directly its destiny. Even the final composition of Rosneft', as confirmed by government resolution 971 in September 1995, was a testament to this observation: thanks to Berezovskiy's earlier-mentioned manipulations, two of the company's best assets were turned over to Sibneft'. This was despite Putilov's agreement with Chernomyrdin – concluded a few weeks earlier – to retain these very assets for Rosneft'.[13] A key reason for Putilov's limited success in stemming the haemorrhage of subsidiaries to other VIOCs was that he was a relative newcomer to the company and its subsidiaries, having worked previously at Uray, a LUKoil subsidiary. He therefore lacked the informal, *de facto* ownership rights wielded by Fomin at Slavneft' or Muravlenko at YUKOS. He could not even aspire to any potential ownership rights or an increase in the value of his shares prior to 1995, in view of the absence of a plan to privatize Rosneft'. Only after the introduction of a new charter for Rosneft' in 1995, which envisaged the divestiture of the state's shareholdings in Rosneft' after three years, did Putilov play a more integral role in the destiny of the company.

Attempted Privatization

The privatization of Rosneft' in 1998, as envisaged by decree 327 of April 1995, appeared to be inevitable. In September 1997, Chernomyrdin signed an order expunging Rosneft' from the list of strategically important enterprises prohibited from privatization. A month later, Potanin withdrew his two-year-long legal battle for ownership over Purneftegaz, a major subsidiary of Rosneft'. Then in December, Yel'tsin rescinded an earlier restriction on foreign investment in oil companies that were to be privatized,[14] thereby encouraging the creation of formidable alliances to bid for Rosneft'. As such, it was widely expected that the auction would be fiercely competitive: indeed, the company's relatively undeveloped oilfields in Sakhalin and the Caspian meant that 'whoever buys Rosneft', buys the future'.[15]

However, despite holding auctions in May, July and October 1998, no bids were received. According to conventional wisdom, the failure to privatize Rosneft' was a result of instability on world financial and oil markets, which in turn affected the investment climate in Russia. For instance, Royal Dutch/Shell explained that its withdrawal from the Rosneft' auction was based on the 'depressed outlook for oil prices, which [did] not justify a bid under current conditions and also the present difficult financial conditions in Russia'.[16] The following section will analyse the extent to which such exogenous factors were responsible for the failed privatization of Rosneft', as well as the roles played by key decision-makers within Russia.

Management conflict

One factor that significantly influenced the process of privatization at Rosneft' was the year-long leadership struggle at the company between its management and board of directors. The conflict originated in late April 1997 when, for reasons that will be explained later, Chernomyrdin dismissed Putilov as president and chairman of Rosneft'. A Rosneft' manager Yuriy Bespalov was named in his place. Many of Putilov's colleagues were outraged with Chernomyrdin's decision since, according to the first vice-president of Rosneft' at that time, Rair Simonyan,

Rosneft' was 'if not the most effective . . . then at least one of two or three best companies in the oil industry'.[17] They suspected that Berezovskiy was behind Putilov's removal, having deployed his tried and tested method of 'privatizing management by removing those who were opposed to him and planting those who could serve his interests'.[18] Indeed, there had been no love lost between Berezovskiy and Putilov ever since the unexpected creation of Sibneft' in late 1995, from entities that Rosneft' believed it owned.

Putilov's colleagues lobbied hard for his reinstatement. In particular, they sought help from the company's banking and foreign partners who were greatly worried about Putilov's removal because 'the name of the head of Rosneft' figured, directly or indirectly, in all of the investment projects of the company'.[19] Less than two weeks later, another government resolution reinstated Putilov as chairman of the board of directors, but left Bespalov in his post as president of Rosneft'. This compromise did little to end the impasse. In November 1997, Putilov's board of directors sent a letter to Chernomyrdin requesting that Bespalov be removed on the basis of the company's worsening performance.[20] On his part, Bespalov accused Putilov of financial impropriety and of trying to engineer the company's takeover by ONEKSIMbank.[21]

The impact of the management conflict on the privatization of Rosneft' was twofold. Firstly, it complicated the preparatory work that needed to be done prior to an auction of the company's shares. A case in point concerned changes to the list of employee-shareholders.[22] During negotiations to corporatize Rosneft' in 1995, it had been agreed that discounted shares would only be distributed to those employees who had joined the company prior to December 1995. In October 1997, however, Bespalov demanded that the approximately 50 new employees whom he had personally recruited during the summer be included as employee-shareholders. This necessitated a legal change in the company's charter, which had to be submitted for approval by several state officials. Bespalov's behaviour was probably a ploy to drag out the preparatory process; indeed, he had earlier rejected a similar appeal by 80 other employees who had joined the company after the December 1995 cut-off date.

The tender for a registrar for Rosneft' was a further example of how preparations for privatization were delayed by management problems.[23] When Putilov was president of Rosneft', the Central Moscow Depository won a tender and was appointed as its registrar. Upon being appointed president, however, Bespalov concluded an agreement with another registrar (the Fund Registration Company) linked to and used by Berezovskiy's other companies. In all likelihood, Bespalov calculated that his action would be opposed by Putilov's board of directors – it duly refused to ratify the said agreement – and would result in the organization of a new tender for a registrar, thereby further delaying the company's privatization. This was because by law, a company's registrar was required to update and release information on current shareholdings prior to any sale.

These delays in making preparations for the privatization of Rosneft' were blamed on Bespalov, who was suspected of attempting to buy time for Berezovskiy to find an alternative partner, following Gazprom's rejection of his entreaties to form an alliance to bid for Rosneft'.[24] At that time, Bespalov's efforts to delay the privatization of Rosneft' did not appear to be fatal to the company's sale. However, with hindsight, it is possible that had preparations for the sale been completed earlier, the company could have been sold at the end of 1997, as initially envisaged by GKI, thereby avoiding the worst of the fallout from the financial and oil price crises in 1998.

Intragovernment strife

As the owner of Rosneft', the government could have resolved the management issue more effectively, for example, by dismissing both Putilov and Bespalov. Instead, the compromise arrangement – dividing the company's top posts between the protagonists – worsened the conflict and further delayed the company's privatization.[25] This apparently irrational reaction by the government is best understood by considering a simultaneous 'game' that was being conducted by the same players at a more strategic level. Specifically, the government allowed the conflict to perpetuate because key representatives in the state apparatus, who were already engaged in a battle for influence with one another, found Rosneft' to be a useful instrument in their struggle. As such, they preferred to retain the status quo and had no incentive to come to an agreement over the sale of Rosneft'.

The intragovernment strife pitted Chernomyrdin, the prime minister, against Chubays and Nemtsov. The latter two had been appointed first deputy prime ministers by Yel'tsin in March 1997, and they controlled the key ministries of Finance and Mintopenergo, respectively. On one level, the rivalry between Chernomyrdin and the 'young reformers' centred on economic issues such as budgetary stabilization, tax collection and reform of the coal industry. At a more intrinsic level, the rivalry was about political power in the post-Yel'tsin era, which both sides felt necessitated control over the country's strategic resources in the energy sector. The 'young reformers' managed to force Gazprom to pay its huge arrears to the budget under threat of asset seizure and appoint a like-minded ally as chairman of the company.[26] For Chernomyrdin, these attempts to undermine Gazprom were perceived as personal attacks on his political interests, since the company – and its financial, media and organizational resources – was supposed to serve as the vehicle for his candidacy in the 2000 presidential elections. Worse still for Chernomyrdin's presidential ambitions was the fact that Nemtsov was widely regarded as Yel'tsin's heir apparent. Other examples of the fight to control key assets included Nemtsov's dismissal of the presidents of state-owned Slavneft' and Transneft' (the operator of the country's pipelines), who were allies of Chernomyrdin, as well as the appointment of Chubays as head of the state-owned electricity company, Unified Energy Systems. The 'young reformers' were apparently trying to 'build their own little oligarchy with the help of administrative or state resources'[27] in order to 'put together an effective rightist bloc'.[28] In any case, by mid-1997, it appeared that the 'key levers of power [had] moved into the hands of a close-knit liberal team' and that 'in terms of influence on the state of affairs in the country, Chubays [had] moved far ahead of Viktor Chernomyrdin'.[29]

The involvement of Chernomyrdin and the 'young reformers' in management issues at Rosneft' reflected this political rivalry. The former initiated the government resolution of April 1997 dismissing Putilov in favour of Bespalov. Noting that some people were trying to 'get the company for free',[30] Nemtsov reinstated Putilov but only as chairman. In an intensification of the dispute and to get Berezovskiy 'knocked out of the battle for Rosneft'',[31] Chubays ordered the seizure and disposal of Berezovskiy's highly prized Omsk refinery, a subsidiary of Sibneft' that owed billions to the budget. Chernomyrdin simply overturned Chubays's order and gave the refinery extra time to repay its debts, thereby confirming suspicions that he was Berezovskiy's 'life buoy at the White House'.[32]

This same power struggle within the government was also mirrored in the privatization plan for Rosneft'. While there was general agreement on privatizing Rosneft', the key actors in this issue – the company's management, key government figures and the business tycoons – were divided among themselves and with each other about whether the government should remain the majority owner. One party in the debate, comprising Berezovskiy, Chernomyrdin and Rem Vyakhirev (Chairman of Gazprom), argued that not more than 51% of the state's shares should be sold since they already had *de facto* control over Gazprom. They were opposed by a second group including Chubays in GKI, Nemtsov in Mintopenergo as well as by Potanin, who backed the divestment of a much-larger stake of 75%-plus-one so as to undermine existing insider control of the company.

Competition among Tycoons

Soon however, changing circumstances resulted in a consensus on the privatization plan. The most significant factor that affected the decision-making environment was the inclusion of foreign partners. In November 1997, the announcement by Royal Dutch/Shell that it was joining forces with Gazprom and LUKoil to bid for Rosneft' was followed, a few days later, by a similar tie up between Potanin's ONEKSIMbank and British Petroleum (BP). These announcements united Russia's previously feuding tycoons and government officials behind the 75%-plus-one option for privatizing Rosneft'. Chernomyrdin and Vyakhirev now concurred with the preferences of the 'young reformers' and Potanin: they calculated that the financial strength of the new Gazprom/LUKoil/Royal Dutch/Shell consortium would give it a greater advantage in any competitive tender for a 75%-plus-one share, whereas a less expensive sale of 50% of Rosneft' would attract more bidders. Berezovskiy, who was unable to entice Western investors to submit a joint bid, continued to be the lone voice arguing that government should retain majority control over the company's shares; this was because he had already 'privatized' the management of Rosneft' thanks to his relationship with Bespalov. The subsequent change of prime minister one week later did not upset the new status quo because Chernomyrdin's successor, Sergey Kiriyenko, was determined to stabilize the deteriorating macroeconomic situation in Russia.

Hence, corporate actors in Russia, such as Gazprom/LUKoil and ONEKSIMbank, were able to manipulate the rules of the game in their favour by introducing new actors. The foreign oil companies made a qualitative difference to the policy process because they provided the bulk of the financing for their respective bid consortiums. As such, their interests had to be taken into account by existing actors. One of these interests was to obtain a 75%-plus-one stake in Rosneft', since this would provide 'more certain control over the company'.[33] A second was to eliminate the proposal for the state to retain a 'golden share' in Rosneft', that is, absolute veto power over management decisions even with a minority stake. Yet another was the appointment of an external, independent, consultant to assess the value of the company. Previously, all such evaluations had been within the purview of the state agency, the RFFI. In any case, the decision of key actors to cooperate on the terms of sale for Rosneft' resulted in the announcement, in March 1998, of an auction for a 75%-plus-one stake at a starting price of US$2.1 billion.

Poor management of Rosneft' sale

However, this consensus was a case of too little, too late, in view of the rapidly deteriorating conditions on the global financial and oil markets. In fact, as explained below, a series of ill-timed and ill-advised actions on the part of the government and the president, between March 1998 and the run-up to the auctions in May and July 1998, only worsened the investment climate in Russia and resulted in the failure to divest Rosneft' from state-ownership.

In the first place, the government's poor management of the Rosneft' sale encouraged greater risk adversity from potential investors. For example, the starting price in the May 1998 auction was set at a level (US$2.1 billion) that greatly exceeded the valuation (US$1.6–1.7 billion) by the external consultant Dresdner Kleinworth Benson.[34] The high asking price reflected the government's desperation to meet budgetary targets following the failure of share divestments in other oil companies in 1997 and 1998 – after all, proceeds from the sale of Rosneft' had already been written into the federal budget for 1998. Another example of poor management of the Rosneft' sale was the statement, in April 1998, by GKI that the government was considering alternative methods of privatizing the company should the May auction fail to attract bidders.[35] GKI therefore 'talked down' the upcoming sale and held open the prospect for investors that the government was prepared to reconsider the sale terms. This encouraged already wary but rational investors to abandon the May auction for Rosneft', in the hope of securing more attractive sale terms for a new auction.

Apart from poor management of the Rosneft' sale, Yel'tsin himself instigated actions that undermined key players in the company's privatization. Specifically, just before the second auction in July, Yel'tsin supported Kiriyenko's attempt to seize Gazprom's assets as a result of tax arrears. Yel'tsin acted out of a desire to appoint a successor who would guarantee him and his family members future immunity from prosecution. In this regard, he felt that Gazprom was the financial and organizational basis for his political rivals, including Chernomyrdin, whom Yel'tsin had dismissed in March 1998 on account of his becoming less beholden to the Russian president.[36] The tax assault on Gazprom was therefore a tool for Yel'tsin to underline his continued control over the political arena.

As for Kiriyenko, his behaviour was neither irrational nor a case of misjudgment – he knew that it would provoke a negative reaction from Gazprom's partners. He remarked, for example, that Royal Dutch/Shell had withdrawn from the Rosneft' bid because his actions had demonstrated that Gazprom 'could not hope for special relations'.[37] The explanation for Kiriyenko's assault on Gazprom lay in the fact that by July 1998, the sale of Rosneft' had lost much of its urgency for Kiriyenko and the 'young reformers', because it was highly likely that the IMF would agree to a bailout package to stabilize the rapidly depreciating ruble. Indeed, Chubays's appointment in June to head Russia's negotiations with the IMF had already led to the release of a tranche of US$670 million that had been agreed upon under an earlier arrangement, but which had been frozen due to poor tax collection targets. Kiriyenko's tax assault on Gazprom was, therefore, motivated by his desire to ingratiate himself with Yel'tsin in order to prolong his tenure as prime minister and thereafter be nominated by Yel'tsin as his successor to the presidency. For critics of Kiriyenko, his actions against Gazprom appeared illogical because 'the success of the country primarily depends on the stability of the government and the welfare of Gazprom'.[38] Kiriyenko, however, possessed 'private information'[39] unavailable to his critics about the impending multibillion dollar loan from the IMF; his assault against Gazprom was therefore based on his calculation that this would not affect the country's financial stability.[40]

External market conditions

To be sure, the Russian government did attempt to assuage doubts about Rosneft' on the part of potential investors. Following the failure of the May tender, it sacked and replaced both the board of directors and management of Rosneft' and lowered the starting price for the July tender to US$1.6 billion. When the July tender again failed to attract any bids, the government appointed external consultants to enhance the financial and commercial attractiveness of Rosneft' ahead of a rescheduled tender in the fall. However, conditions on the global financial and oil markets were deteriorating rapidly. The financial crisis in Asia, which began in October 1997, exacted tolls on stock markets all over the world, particularly those in emerging markets such as Russia. For example, the value of the Russian stock market, which had grown by 88% between 1996 and 1997, fell by more than 50% between January and May 1998.[41] The ruble came under heavy pressure as concerns about Russia's ability to service its huge domestic and foreign debts intensified. Finally, on 17 August 1998, the ruble was devalued.[42]

As for the price of oil on world markets, it continued to deteriorate from US$19 per barrel in November 1997, when foreign oil companies announced their intention to participate in the sale of Rosneft', to US$13 per barrel in May 1998, when the first tender was held, and to merely US$11 per barrel when Royal Dutch/Shell and BP announced their withdrawal from the auction in July.[43] Hence, by mid-1998, the purchase of an oil company was not only perceived as a suspect investment because of its declining value, but oil companies interested to bid for Rosneft' also found it increasingly difficult to borrow money from financial markets because oil was suddenly a less secure form of collateral. Following the collapse of the ruble, Yel'tsin sacked Kiriyenko. His successor, Yevgeniy Primakov, finally called off the sale of Rosneft' in September 1998.

It should be clear from the preceding discussion that the repeated failures to privatize Rosneft' were not fundamentally caused by the dual crises in financial and oil markets. The deteriorating market conditions only aggravated the already pessimistic perceptions of Royal Dutch/Shell and BP concerning the wisdom of purchasing a company with pre-existing and unresolved issues. Neither was the fiasco over the privatization of Rosneft' shaped primarily by competition between the tycoons to acquire the company,[44] although it certainly fuelled the bitter management feud and raised more questions about the company's financial health and standards of professionalism. Such explanations erroneously downplay the role of the state in the whole affair, namely, the willingness of the president and key government officials to use the privatization of Rosneft' as a weapon in their struggle for power. The attractiveness of the company as a target of acquisition was, therefore, heavily compromised even before any auction took place. It is conceivable that Rosneft' could have been privatized by late 1997 or early 1998 – thereby avoiding the impact of the financial and oil crises – if consensus about the size of the share divestment had been reached earlier. As it turned out, the intentional delay to withhold cooperation in crafting such a consensus among the managers of Rosneft', key government officials and leading tycoons proved to be fatal for the company's privatization.

The new look Rosneft': Creeping privatization?

In October 1998, Sergey Bogdanchikov, a relatively obscure career oilman into his fifth year as general manager of Sakhalinmorneftegaz, a Rosneft' subsidiary, was appointed president of Rosneft'. In many respects, he was in the proverbial right place at the right time: Bogdanchikov was overshadowed by more illustrious candidates for the post but found favour with the then prime minister Primakov, precisely because he was not closely aligned with any existing financial or political groups.[45] He quickly echoed Primakov's sentiment by stating that it would be 'better to forget about the privatization of Rosneft' for the time being', and that his priority was 'to integrate and strengthen the company'.[46] In an interview in 2003, he reiterated his conviction that any privatization of Rosneft' was 'premature' despite the company's much-improved financial stability.[47] Yet, by 2006, he was at the forefront of a campaign that divested 14.9% of the company. Why and how this policy change occurred is the subject of the following section.

From a rejection of privatization . . .

A cursory glance at any list of richest Russians suggests that the surest route to wealth involved controlling the cash flow of a soon-to-be privatized company in the natural resource sector either directly as a member of senior management, or indirectly as a key shareholder. For Bogdanchikov, this meant that his priority was to maximize his unexpected tenure at Rosneft' so as to nudge it towards the kind of privatization from which he would benefit financially. Therefore, like his predecessors, his first order of business was to ensure the survival of Rosneft' itself as a company. Between 1998 and 2000, companies affiliated with Putilov had managed to seize temporary control of 38% of Purneftegaz, which accounts for 60% of the total output of Rosneft', and to dilute the assets of other oil-producing subsidiaries.[48] This underlined the crucial need to consolidate control over the company's subsidiaries even before considering privatization. In this connection, Rosneft' spent at least US$500 million buying up shares held by minority investors in its subsidiaries so as to increase stakes held by the holding company from 38% to more than 51%. Consequently, Bogdanchikov managed to minimize avenues available to potential outside owners to control financial flows, shares, debts or management staff of subsidiary companies as a back-door route to the acquisition of shares in and control rights over the parent company Rosneft'. At the same time, he also gained 'a free hand in disposing of their [the subsidiaries'] financial and commodity flows'.[49]

A second, and more significant, constraint on the privatization of Rosneft' at this time concerned the policy choices and behaviour of Putin. As previously noted, Putin's highest-order preference is his own political longevity. During his first term, this implied maximizing his tenure as president through re-election, while in his second term, this preference is reflected in efforts to 'put into practice the power reproduction process – the search for and choice of a successor, or extension of Putin's presence' on the political scene.[50] With an eye on his re-election in 2004, Putin had to make himself and his policies relevant to most of Russia's political constituencies.

To appeal to the war-weary population, state-owned Rosneft' was deployed to pacify and 'normalize' Chechnya. In January 2000, Putin instructed Rosneft' to take charge of all rehabilitation work in the Chechen oil industry, and in November, the Grozneftegaz company – jointly owned by Rosneft' (51%) and the Chechen administration (49%) – was created by the government for this purpose.[51] Rosneft' acts as a surrogate for Putin's policy in Chechnya: in return for minimizing incidences of illegal siphoning of oil and reducing attacks on oil infrastructure by rival Chechen clans, the pro-Russian Chechen administration apparently receives part of the proceeds from the company's monopoly on oil sales there. This arrangement allowed Putin and the pro-Russian Chechen administration to claim that Chechnya is relatively autonomous from Moscow, hence stabilizing the situation in the republic.

To appeal to Russia's political elites, Putin retained the services of Yel'tsin-era administrators such as Voloshin and Chubays even as he forced into exile other Yel'tsin-era stalwarts such as Berezovskiy. Putin also realized that he had little to gain from Russia's financial elites by privatizing Rosneft'. As explained in Chapter 4, Russian tycoons then were more focused on consolidating their existing business empires in the wake of the country's financial crisis of August 1998; they were willing to spend money on acquisitions only if they were assured of becoming majority shareholders. However, GKI was considering selling only a minority stake in Rosneft'.

In short, therefore, neither Bogdanchikov's management team at Rosneft', nor Putin, nor Russia's business tycoons felt that the sale of Rosneft' would advance their respective self-interests. For Bogdanchikov, it is to continue as head of Rosneft' for as long as possible. For Putin, it was to maximize his political longevity as president and 'king-maker'. As for the oligarchs, it was simply the preservation of their business empires and wealth. During Putin's first term, the retention of Rosneft' as a state-owned enterprise was the vehicle through which these different interests were protected. Following Putin's electoral victory in March 2004, such interests remained constant. What changed, however, was the realization that in the run-up to a post-Putin era, the interests of all concerned parties would be best secured by cooperating on implementing a radically new strategy – the partial divestment of Rosneft' from state-ownership.

... To an agreement on partial privatization

The imminent return of Yuganskneftegaz to state-ownership as a result of the YUKOS affair and the continuing upward spiral in global energy prices provided the impetus for Putin to realize his ambition, highlighted in the previous chapter, of enhancing state control over Russia's energy sector. Towards this end, in September 2004, Putin endorsed a proposal to exchange 100% of Rosneft' for an additional 10.7% of Gazprom's shares in order to increase the state's stake in the latter from 38.37% to over 50%. This would give the state majority control over an entity accounting for 20% of the world's annual gas production, 35% of global gas reserves, 20% of federal budget revenues and up to 8% of Russia's GDP.[52] As a consolation, Bogdanchikov was named the head of a new oil subsidiary within Gazprom that would consolidate the company's oil assets, among which Yuganskneftegaz and Rosneft' were expected to figure.

Bogdanchikov, however, had little to gain from cooperating with the proposed merger, since his new position as head of a mere subsidiary would have entailed a huge loss of authority and freedom of manoeuvre. For Sechin, the chairman of Rosneft', the company has been an important power and monetary base with which to establish unquestioned supremacy for himself within the presidential administration. Indeed, 'the profits of Rosneft' . . . are almost seven times greater than the profits of Transnefteprodukt'[53] where a rival from the presidential administration, Vladislav Surkov, serves as chairman. Also, the 'disappearance of Rosneft' into Gazprom would mean the unchallenged preeminence of Dmitriy Medvedev':[54] the latter is chairman of Gazprom and as head, of the presidential administration, is Sechin's boss. This personal rivalry is made worse by the fact that Sechin and Medvedev represent different 'factions' within the administration, namely, the *siloviki* (from the security and intelligence services) and 'petersburgers' (comprising mainly economists and jurists), respectively. Bogdanchikov and Sechin chose to react to this threat to their interests with a series of counterproposals that ranged from retaining Rosneft' as a state-owned entity to its partial privatization.

In an early attempt to alter the terms of the Gazprom–Rosneft' merger, Bogdanchikov suggested that the state should retain part of Rosneft', since 75–80% of the company's shares would be enough to acquire 10.7% of Gazprom's shares. He appointed an investment bank, which valued Rosneft' at US$7–8.5 billion or more than enough to acquire 10–14% of Gazprom. This would have allowed him to retain his position as head of a rump Rosneft'. However, Medvedev and Aleksey Miller, president of Gazprom, were equally determined to absorb all of Rosneft' within Gazprom, since an enlarged company would be in their pecuniary interests.[55] They hired an alternative investment bank, which duly reported that Rosneft' was worth US$5–5.7 million or less than the 10.7% stake in Gazprom.[56]

However, just as institutions can constrain an actor's scope of action, it can also open up new opportunities for him. In this connection, Bogdanchikov and Sechin took advantage of the new rules of the game created unwittingly by YUKOS: the company obtained a temporary injunction against Gazprom and its subsidiaries, threatening a seizure of their assets in the United States should they participate in the December 2004 auction for Yuganskneftegaz.[57] Subsequently, Rosneft', which at that time had no US-based assets, participated in and won ownership of Yuganskneftegaz. This in turn presented Bogdanchikov with the leverage to eventually extricate Rosneft' from the merger with Gazprom. As he himself noted, 'whereas in November [2004] Rosneft' presumed that in 2005 production would amount to approximately 25 million tons of oil, now it is a company that will produce 78–79 million tons per year. Objectively speaking, this is an absolutely different asset. That is why the previously envisioned scheme should be transformed'.[58]

Bogdanchikov, however, was not simply a lucky manager who took advantage of the opportunity presented by YUKOS's injunction to create a more powerful Rosneft'. He also actively and successfully persuaded the Kremlin to adopt two key decisions; the first being that Rosneft' would be the most suitable owner of Yuganskneftegaz and other assets of YUKOS. Indeed, most oil analysts and political commentators had expected Surgutneftegaz to win the December 2004 auction in view of its loyalty to the Kremlin, its large cash reserves and the proximity of its oilfields to Yuganskneftegaz.[59]

Secondly, Bogdanchikov convinced the Kremlin that his alternative scheme – a partial sale of shares in Rosneft' through an Initial Public Offer (or IPO) – would be more advantageous than a Gazprom–Rosneft' merger, particularly in terms of advancing Putin's self-interest. Retaining Rosneft' as a separate entity means that there are two national champions, instead of just one, in the energy sector, each run by Putin's trusted associates representing different 'factions'. This was clearly a more prudent coalition management tactic since merging Rosneft' and Yuganskneftegaz into Gazprom could have alienated Sechin and his *siloviki* and increased Gazprom's potential as the sole, competing centre of power to the presidency. Putin also ensured that he was indispensable as both sides compete to influence his choice of successor for the 2008 elections. Furthermore, there was the populist appeal of giving the Russian public a chance to own one of the most valuable Russian companies through an IPO of Rosneft': this enhanced Putin's own political legacy and the popularity of his chosen successor, particularly since the public was largely excluded from all previous privatizations of Russia's oil and gas companies.[60] In comparison, the Gazprom–Rosneft' merger necessitated only an internal share swap between the two companies.

For Putin, Bogdanchikov's proposal was also attractive because it retained one of the key features of the Gazprom–Rosneft' merger, namely, that it gave the Kremlin – rather than the Duma – control over the entire process. This was because the money to purchase a 10.7% stake in Gazprom was repaid through an IPO of Rosneft', instead of the usual procedure of setting aside federal funds in an amended budget, a process that would have required several readings in the Duma followed by a vote. Proceeds from the sale of Rosneft' therefore accrued not to the state but to Rosneftegaz, a specially created state-owned investment vehicle to administer the state's remaining stake in Rosneft' and its 10.7% share in Gazprom.[61] By selecting the IPO option over the conventional privatization of state property as managed by GKI, the Kremlin and people within Putin's administration 'are personally the beneficiaries and can divide the money among themselves without being accountable to anyone'.[62] They include the head of Rosneftegaz (Nikolay Borisenko, a former first vice-president of Rosneft') and the board of directors of Rosneftegaz, notably Bogdanchikov and Sechin.

The determination shown by Bogdanchikov and Sechin to privatize Rosneft' stemmed not from any deep-seated convictions about the merits of privatization *per se*, but from a calculation of how a partially privatized Rosneft' under their control would serve their pecuniary interests. Indeed, one source claimed that the intention was for Rosneft' to be 'corporatised before Putin leaves office so that Putin's successor cannot fire Bogdanchikov, at least not easily',[63] particularly since the company's owner, Rosneftegaz, is favourably disposed towards him and his allies. To further protect his position, Bogdanchikov insisted that no single shareholder, including institutional investors, be allowed to own more than 2% of shares in Rosneft'.[64] Bogdanchikov's concern with safeguarding his tenure at Rosneft' from shifting political winds is also reflected in an attempt, albeit unsuccessful, to sell up to 49% of the company's shares in the IPO. However, the low take-up rate by foreign investors wary of possible legal claims filed by YUKOS over Yuganskneftegaz forced him to reduce the stake on offer to 30% and finally to 14.9% or just enough to repay the loan for 10.7% of Gazprom's shares.[65] In any case, Bogdanchikov's focus on protecting his position at Rosneft', while illustrative of rational choice behaviour, may also have been fuelled by past experience in 1999, when there had been rumours that Putin, then prime minister, was about to dismiss Bogdanchikov as head of Rosneft'.[66]

Turning to Medvedev and Miller at Gazprom, they would clearly have preferred to proceed with the merger. Nonetheless, their agreement to call off the merger reflected the 'satisficing' behaviour of rational individuals who accepted a lower-order preference because it embraced a minimum level of their self-interest. In this light, Putin's approval of the sale of a 72.6% stake in Sibneft' to Gazprom in September 2005, which was meant to appease Medvedev's 'petersburgers' over the sale of Yuganskneftegaz to Rosneft', made ending the merger more palatable for Gazprom since the company would increase its influence in the lucrative oil production sector from 3% to 10%. Although this is a less optimal share than control over Rosneft' and Yuganskneftegaz would have bequeathed (19%), it is, nevertheless, a significant improvement over Gazprom's hitherto negligible role in the oil sector.[67]

As for the country's tycoons, they had little incentive to oppose Putin's decision to partly privatize Rosneft'. The YUKOS affair had driven home the high price of opposition to the Kremlin, and Russian tycoons, including Abramovich, were quick to demonstrate their loyalty by participating in the IPO.[68] This was despite the fact that the amount of shares they acquired were too small to be of significant investment value, and that the price per share was overvalued. In this connection, participation in the IPO was more an exercise in 'relationship building' with the Russian authorities.[69]

Hence, with the approval of the major actors as outlined above, the merger between Gazprom and Rosneft' was officially called off in May 2005. In June 2006, 14.9% of Rosneft' was sold for US$10.4 billion in an IPO, and following a share consolidation within subsidiaries of Rosneft', the state today owns 75.16% of Rosneft' through Rosneftegaz.

Observations about the case study of Rosneft'

Slavneft' versus Rosneft'

ONAKO and Slavneft' were different from Rosneft' as objects of privatization, despite the fact that all three companies operated under the same set of political conditions during Putin's first term. On the one hand, Putin decreed in September 2000 that Rosneft' would manage the government's share in production-sharing ventures in Sakhalin.[70] On the other, he refrained from commenting on the fates of ONAKO and Slavneft', both of which were eventually sold to private owners. One plausible explanation is that the original conception of Rosneft' as a quasi-national oil company provided it with a 'safeguard' against privatization at that time: in other words, the development of Rosneft' appeared to be a triumph of structure over agency.

To be sure, Rosneft', more than any other oil company in Russia, has long been regarded as the country's *de facto* national oil company. This was partly due to the provisions in its founding and revised charters, which endowed Rosneft' with some functions that are usually the purview of a national oil company. The perception of Rosneft' as a *de facto* national oil company was also encouraged by heads of Mintopenergo for reasons of job security. Nevertheless this legacy and perception of Rosneft' as a national oil company became part of the policymaking environment only because influential actors calculated that its use or advocacy was useful for their own personal ends. When this same legacy was deemed to be inconvenient by key actors – such as during attempts to privatize Rosneft' in 1998 or the plan in 2004 to render Rosneft' an oil subsidiary of Gazprom – it was quietly ignored. Historical institutionalism thus appears to be a less convincing framework, compared to rational choice institutionalism, with which to explain the different privatization histories of Slavneft' and Rosneft'.

Self-interest disguised as state interest

The analyses of the YUKOS affair and policy change over Rosneft' reinforce the image of Putin as an intendedly rational politician seeking to maximize his political longevity, in terms of consolidating his hold on the presidency and retaining enough influence to subsequently choose his successor. Putin's self-serving motives were clearly reinforced by pre-existing informal institutions or parameters and his experience as a KGB officer, such as his deep belief concerning the leading role of the state, as discussed earlier. Nevertheless the contention is that self-interest, rather than ideas or beliefs, is the primary (albeit not the only) determinant of Putin's behaviour. This is the most persuasive explanation for the sale of Slavneft' in 2002, which occurred despite the contents of his 1999 dissertation. The case study of Rosneft' has further demonstrated that notwithstanding his professed faith in the state, Putin repeatedly favours only that part of the state which he directly controls – the presidential administration – at the expense of the government and wider bureaucratic agencies.

How else to explain the fact that the partial privatization of Rosneft' deliberately bypassed parliamentary procedures for sale of state property? How else to explain the fact that the state budget was not at all compensated for the sale of state property, an outcome that has been criticized as theft on a scale larger than the loans-for-shares auction?[71] How else to explain the fact that Rosneftegaz has not yet been liquidated following the repayment of credit for additional Gazprom shares, and that its assets – Rosneft' and the 10.7% of Gazprom – have yet to be transferred to GKI for management as per the original plan? In the latter's case, this is because Rosneftegaz is now viewed as a useful vehicle with which to bypass government oversight of future privatizations, particularly with Sechin on its board of directors.[72] How else, also, to explain Putin's 'dual oversight' system – that is, appointing presidential administrators as chairmen of state-owned companies – which implies a lack of trust in state-appointed chief executives of those companies?[73]

The primacy of self-interest as a key determinant in the decision-making process was also manifested in the rivalry between the leaders of Gazprom and Rosneft'. The tussle over the future of Rosneft' was not simply driven by the ideational goal of giving the state more control over the commanding heights of the economy in order to put right the controversy over privatization in the past.[74] It was also about a struggle between two groups of people intent on maximizing their pecuniary preferences, who occupy key positions within the state apparatus and state enterprises and who desire to claim these commanding heights as their very own.[75] Indeed, ministers at that time who were board members of Gazprom or Rosneft' took up their respective company's cause in the debacle. For example, Sergey Oganesyan, head of the Federal Energy Agency and deputy chairman of Rosneft', argued that 'the shares of Yuganskneftegaz must remain with Rosneft' . . . [in the light of] special, technological special features and differences in oil and gas production'.[76] In contrast, Viktor Khristenko, the Minister of Industry and Energy and Gazprom's board member, stated that Yuganskneftegaz would be separated from the composition of Rosneft' and that the latter would be merged with Gazprom as per the original plan.[77] Fellow board member and Minister of Economic Development, German Gref, echoed Khristenko's comments.[78]

It therefore appears that references to 'state interests' or 'national interests' made by public sector actors have limited utility in contributing to our understanding of the decision-making environment in Russia, particularly when it is applied to policies within the oil industry. It is only by desegregating components that make up the state – the presidency, federal ministries, regional authorities, state-owned enterprises – that a complex web of bureaucratic, ideological and personal interests become apparent when explaining behaviour that is often cloaked in patriotism. In this regard, it is difficult not to agree with Khodorkovskiy's assessment that for Bogdanchikov, Sechin and others, the Russian state is merely a 'mechanism of promoting their personal interests' and a 'hostage to the interests of certain individuals wielding the power of state officials'.[79]

6
Conclusion

The focus of this book has been the dynamic relationship between business and the state with respect to the policy of privatization in the Russian oil industry. In this connection, two key questions were posed in the opening chapter. The first concerns the perennial Russian fascination with '*kto kogo*' or 'who does what to whom', and specifically, whether business–state relations are more accurately characterized as 'capture' or exchange. The second concerns an often overlooked segment of the business community, the Soviet-era general managers of oil companies. To what extent was their role in policymaking determined by institutional incentives in the post-Soviet era, rather than by values and personal networks inherited from their *nomenklatura* status? This chapter will address these issues by taking into account the foregoing analyses of the initiation of oil privatization and the specific case studies of YUKOS, Slavneft' and Rosneft'. It will then consider the utility of the rational choice institutional approach as applied to policymaking on oil privatization.

The state and the oil industry: A dynamic relationship

This book has argued that the conventional wisdom that tycoon-owned corporations have taken advantage of the weak state and privatized policymaking in Russia is simplistic and ignores the shifting dynamics of the business–state relationship. Prior to 1995, developments in the oil industry were shaped by the preferences of the *nomenklatura* general managers of leading oil enterprises and key officials from various state bureaucracies and the government. They were eager to cooperate with one another to advance their respective preferences. These included reducing the role of the state in the economy by increasing the spread of private ownership (the 'young reformers'), retaining the presidency by entrenching market and democratic reforms (Yel'tsin), self-enrichment through asset-stripping (the *nomenklatura* managers) and increasing state control over the oil industry for reasons of bureaucratic survival (Mintopenergo). As a consequence, they became stakeholders in the newly privatized oil industry, institutionalizing their gains from cooperation through decree 1403 and subsequent resolutions that established corporatized oil companies.

During the second half of the 1990s, tycoon-owned companies began to challenge the prominent role played by Soviet-era general managers in privatizing the oil industry.[1] In return for 'satisficing' the political and economic interests of Yel'tsin and the 'young reformers', the tycoons were allowed to pursue their commercial interests by acquiring a limited number of the most lucrative companies. Nevertheless the extent of success in divesting a particular oil company was very much dependent upon the perception among the general managers, tycoons and state officials about the gains from cooperating on privatization.

Take the case of YUKOS. Yel'tsin and the 'young reformers' were eager to sell the state's equity in the company, while tycoons such as Khodorkovskiy and Fridman were equally keen to become its new owners. Muravlenko agreed to support the sale because he understood that his pecuniary interests were better served by cooperating, rather than 'defecting'. He knew that he was not as capable as Alekperov or Bogdanov and was at risk of being dismissed by the state. Hence, cooperating while he was still general manager, and therefore had some bargaining power, was preferable to any future uncertainty about his position. With Muravlenko, GKI, the Ministry of Finance and the Central Bank behind Khodorkovskiy's bid, and with Khodorkovskiy himself acting to undercut Fridman's rival consortium, Menatep won control of YUKOS.

In the case of Slavneft', the breakdown in cooperative relations within the triangular relationship was the main determinant of the failure to privatize the company in 1997–1998. The 'young reformers' were keen to sell the state's 19.68% stake in the company but Fomin agreed to support the sale only if his preferred consortium of foreign investment banks could participate, albeit indirectly. As for the tycoons, they were against the sale because the stake on offer was too small to secure a blocking vote, and they could not hope to compete with the terms offered by Fomin's consortium; they hence chose to 'defect' from cooperation and got the sale annulled and postponed. In contrast, Slavneft' was successfully privatized at the end of 2002 because the key actors were in agreement about the sale. Sibneft' and TNK, who had a close ally in the new president of Slavneft', launched a joint bid to overwhelm the resources of the rival from Mezhprombank. They found support from Putin, who agreed to the sale in exchange for political capital.

Under Putin, the state slowly emerged as the dominant actor within the business–state relationship, which explains why very few actors dared to withhold their cooperation from the Kremlin. Nevertheless there were still attempts by business leaders to craft new rules of the game that advanced their own self-interests: for reasons that were discussed earlier, Khodorkovskiy failed in his endeavour to hold on to YUKOS while Bogdanchikov succeeded in partially privatizing Rosneft'.

It is this constantly shifting balance of power between business leaders (tycoons and *nomenklatura* general managers) and the state, as well as the deliberate choices made by these actors to cooperate or 'defect' from the evolving rules of the game in order to redress the balance to their own advantage, which has determined the scope and pace of privatization in the Russian oil industry. Such choices were not derived exclusively from contemporary situational considerations, but were sometimes also influenced by legacies and past experience. In other words, a full appreciation of the policymaking process must take into account the intendedly rational individual and his choices, existing institutional incentives and constraints as well as the legacies of the past, as this book has tried to do.

'Capture' or Exchange?

As previously noted, there is a debate about whether business has 'captured' or 'privatized' the state, or whether representatives of the state willingly colluded with big business in order to achieve their own, separate and specific aims. Since 2003, this debate may be expanded to include the extent to which the state has 'tamed' business. In any case, the central question is whether decisions concerning oil privatization have benefited only one party or whether there has been a two-way flow of advantages. The preceding chapters strongly suggest that mutual exchange is a more accurate characterization of business–state relations. As noted by Timothy Frye 'that powerful firms are able to extract resources from the state will hardly come as a surprise to observers of the post-communist world . . . That state officials are able to extract resources from business elites, however, has been less well recognised'.[2]

Three major levers were used to extract resources and appeal to the self-interests of state officials and business. One concerned the use of privatization (or de-privatization) as a coalition management strategy to consolidate the power of the incumbent president and enhance his political longevity. Peregudov has noted that while 'it would be erroneous to associate all instances of redistribution of property solely with the presidency . . . the fact that the largest transactions would be impossible without 'connections' on this level is not subject to doubt'.[3] Both Yel'tsin and Putin faced constraints on their authority upon ascending to the presidency. In Yel'tsin's case, he was hamstrung by the fact that the existing balance of social forces in Russia was untouched, so that 'no enterprise directors, no ministerial officials were removed from office, no accounts were seized . . . official positions, funds and connections remained intact'.[4] As for Putin, he was constrained by the inheritance of personnel from Yel'tsin's entourage, including Voloshin and Kas'yanov, the informal norms that had been established with regard to business–state relations, as well as the lack of an independent power base of his own due to his sudden meteoric rise.

To overcome these formal and informal institutional constraints on power, both Yel'tsin and Putin used privatization, particularly in the oil industry, to create coalitions with a direct stake in supporting and enhancing their political longevity and relevance. For example, the corporatization of oil companies through decree 1403 split the ranks of the *nomenklatura* managers and hence reduced the effectiveness of the Communist-led opposition to economic reforms and Yel'tsin's authority. Later, Yel'tsin used loans-for-shares to replace this earlier group of supporters with a new coalition of stakeholders represented by tycoons from the banking sector. In Putin's case, privatization in the oil industry was initially used to encourage Yel'tsin-era tycoons to continue supporting him as president. Subsequently, he agreed to the de-privatization of Yuganskneftegaz and Sibneft', as well as the Rosneft' IPO, to strengthen his own coalition of supporters among the *siloviki*, petersburgers and their business allies. At the same time, he also allowed the business empires of selected Yel'tsin-era tycoons – namely, Abramovich, Fridman and Potanin – to flourish.[5]

For those businessmen privileged enough to be included in these political coalitions at one time or the other, they benefited materially from control over the oil company's cash flows and assets and an expansion of their business empires, as highlighted in the case studies. In short, the use of privatization as a lever of resource extraction benefited both business and presidents.

The second major lever that was deployed with respect to oil privatization was the opportunistic use of policy windows. A policy window opens only temporarily and sporadically and is 'an opportunity for advocates of proposals to push their pet solutions'.[6] In this connection, the acceptance of corporatized VIOCs as the new basis for the post-Soviet oil industry was the result of a policy window that was exploited by Alekperov. As noted earlier, VIOCs represented an international trend in best practices that was prevalent in the late 1980s and early 1990s. The window offered by this trend also coincided with the systemic change in Russia and the appointment of 'young reformers' in the government, all of which gave Alekperov the opportunity to seize the initiative. Whether or not the VIOC trend on its own would have been sufficient to open the policy window towards restructuring the oil industry, if the Soviet Union had not collapsed, is debatable.[7] In any case, Alekperov was rewarded for his initiative with his own oil company, LUKoil; as a result, the GKI under Chubays claimed its first victory in efforts to depoliticize political control over the economy while simultaneously creating a political coalition in support of Yel'tsin and privatization.

A further example of the successful use of a policy window relates to the impetus for the creation of Slavneft' as a Russian–Belarussian oil company. As explained in Chapter 4, between 1993 and 1994 there was a shift in the national mood towards greater activism in the former Soviet republics. As the policy entrepreneur in this case, Fomin persuaded Chernomyrdin that a jointly owned oil company was in Russia's economic and strategic interests. Fomin's integral role in the process should not be underestimated: even Chernomyrdin acknowledged that 'Fomin was involved right from the beginning'.[8] Like Alekperov, Fomin was rewarded for his initiative by his appointment as president of Slavneft'. For Chernomyrdin's government, majority state-ownership of Slavneft' was a useful lever with which to retain and extend its influence in Belarus.

However, not all opportunities presented by policy windows were grasped successfully. In the case of Gazprom, it was presented with a golden opportunity, in the wake of the YUKOS affair, to emerge as the country's pre-eminent oil and gas monopoly incorporating Yuganskneftegaz and Rosneft'. However, rearguard action by YUKOS and Rosneft' put paid to its ambition and slammed shut this brief policy window.

Finally, policy venues were used as another lever by the state and oil-related business to extract resources from each other. A policy venue identifies which institutions in society will be granted jurisdiction over particular issues. Its significance lies in the fact that '[e]ach venue carries with it a decisional bias ... When the venue of a public policy changes, as often occurs over time, those who previously dominated the policy process may find themselves in the minority, and erstwhile losers may be transformed into winners'.[9] In this respect, a policy outsider has the incentive to appeal to actors not currently involved in a policy debate, in the hope of using their participation and support to transform his own losing position into a winning one.

A case in point concerned Gutseriev's decision to involve the Duma in the privatization of Slavneft'. Hitherto, all privatization-related legislation had been introduced by presidential decrees to circumvent expected opposition to such plans by the Communist-dominated Duma. Gutseriev used his influence as a former deputy speaker of the Duma to include the latter as a political actor. The result was that the Duma prohibited the sale of state-owned shares in large enterprises in 2001, including Slavneft'. Gutseriev also persuaded the Duma to instruct the Audit Chamber to investigate Al'fa's purchase of TNK as a way of putting pressure on Al'fa to back away from interfering in Slavneft'. By skillfully shifting the policy venue to the Duma, Gutseriev managed to extract the 'administrative resource' of the state for use against Al'fa, which was trying to increase its influence within Slavneft' at Gutseriev's expense.

Putin was similarly adept at making use of a change in policy venue to extract resources from business. Under Yel'tsin, there were multiple points of access within the presidency and the state machinery to influence policymaking. Khodorkovskiy, for instance, obtained the support of a wide array of state institutions. Today, however, the presidential administration has become the only policy venue of consequence, since Putin 'drained all power out of the formal institutions of government'.[10] This monopoly of access implies that in order to thrive, business is obliged to repeatedly demonstrate its loyalty to the president. For instance, the Russian partners of TNK–BP – Fridman's Al'fa group and Access Industries/Renova owned by Viktor Veksel'berg – are keen to obtain the Kremlin's blessing to sell their 50% stake in the oil company; this is an attempt to cash out ahead of the uncertainty that the 2008 presidential elections may introduce into the political and business environments.[11] Towards this end, Veksel'berg spent around US$100 million in February 2004 to purchase a collection of Fabergé eggs commissioned by the last Russian tsar and displayed them in Russia for the first time since 1917. Similarly, Al'fa's legal suit against the *Kommersant'* group owned by Berezovskiy reflected 'a desire to pick up points with the authorities by harassing the independent media'[12] because the 'destruction or sale of one of the few remaining major independent mass-media outlets in Russia would nicely decorate the list of services performed by this businessman and his partners for the authorities'.[13] In this connection, big business in Russia no longer represents 'countervailing power',[14] but is instead 'rolling over to show their loyalty to the state'.[15]

'Red Directors'

It should be clear from the foregoing chapters that as a group, the *nomenklatura* general managers of oil companies played an integral role in decisions to create, corporatize and privatize their companies. However, their individual contributions differed. Some of them, such as Alekperov and Fomin, were policy entrepreneurs. Others, including Muravlenko and Bogdanov, chose to emulate Alekperov after he had made the first move to lobby for the establishment of corporatized VIOCs. Their continued success in operating as 'oil generals' within the new market-oriented environment also varied enormously. For instance, Alekperov and Bogdanov continue to own and manage their companies successfully, Putilov and Fomin were dismissed in 1998, while Muravlenko sold his ownership rights to an industry outsider, Khodorkovskiy. In other words, the personal connections and networks that these *nomenklatura* managers brought with them into the post-Soviet era were necessary in the beginning but were insufficient guarantees of continued success. Putilov and Fomin fell victims to high politics, while Muravlenko's paternalistic and socially oriented managerial style was found wanting in the new market-based economy.

A related issue here is the extent to which such *nomenklatura* networks account for the different development strategies of companies owned by Soviet-era oilmen and by 'new Russian' tycoons. According to Clifford Gaddy and Barry Ickes, '[w]hat kind of relations, with whom, how solid they are' constitute 'relational capital', a vital asset under the Soviet system embodied in the general manager of a large enterprise.[16] They argue that the higher levels of investment by LUKoil and Surgutneftegaz, compared to YUKOS and Sibneft', are due to the fact that career oilmen feel more secure about their property rights thanks to their special relationships with state officials, which are in turn a reflection of long-established relational capital.[17] Such oilmen are therefore less vulnerable to any arbitrary action, for instance, property expropriation by the state. Moreover, it is claimed that while the stock of relational capital possessed by Alekperov and Bogdanov remains high, that held by Khodorkovskiy had been depreciating over the years. As a result, Khodorkovskiy rationally calculated that quickly improving efficiency and transparency at YUKOS would be a way of protecting his position, since it would enhance the company's value to outsiders, along with his own property rights.[18]

However, Alekperov's relational capital did not protect LUKoil from being one of the first companies, under Putin, to be charged with tax evasion in July 2000.[19] Neither have Bogdanov's good relations with the state put paid to speculation of its takeover by Gazprom or Rosneft'.[20] In this connection, it is suggested that the rational choice institutional approach may serve as an alternative explanation of the divergent development strategies of companies owned by oilmen and tycoons. Surgutneftegaz is clearly the industry leader when it comes to investment, in terms of absolute value as well as per barrel of output; in 2000, for instance, upstream investment by Surgutneftegaz was US$3.75 per barrel of output compared with US$1.11 for YUKOS and US$1.61 for Sibneft'.[21] This may be attributed to the undisputed fact that the oil fields owned by Surgutneftegaz are relatively older and hence more depleted, thereby obliging the company to drill more extensively than YUKOS and Sibneft' to recover the oil. In other words, Surgutneftegaz's high levels of investment may be a calculated and deliberate response to the situational constraint of ageing oil fields, rather than the result of a perception of greater security of property rights thanks to a high stock of historically acquired relational capital.

Moreover, the choice of different business models of management may have played a role in structuring decisions about operational matters.[22] It is the pro-investor business model chosen by YUKOS and Sibneft', rather than their relatively more tenuous claim on property rights and relational capital, that forces these companies to concentrate on investment returns and reduce spending on social benefits. Take the issue of equity held by top management. Prior to the YUKOS affair, Khodorkovskiy owned 59.5% of the equity of Menatep (the holding company for YUKOS). Likewise, Abramovich and a core group of managers control 88% of Sibneft'. As for Alekperov, he directly owns a 10% stake in LUKoil.[23] They therefore benefited directly from an appreciation in the company's sharemarket value.

In the case of Surgutneftegaz, the company largely owns itself through its Pension Fund. Bogdanov and his top managers have voting control in the company and administer the decision-making power in Surgutneftegaz, but they are not the ultimate or direct beneficiary of the company's equity appreciation. As a result, Surgutneftegaz's management are less motivated than owner-managers in YUKOS, Sibneft' or LUKoil to increase the market value of the company's shares. This, in turn, means that Bogdanov is relatively less concerned with indicators and practices that are highly valued by the market, such as transparency and asset efficiency. So, even though Surgutneftegaz spends more on drilling and replacement of upstream infrastructure, the returns on such investments are relatively poor: Surgutneftegaz's wells were 40–55% less productive than those owned by YUKOS and Sibneft'.[24]

Another way in which the choice of a business model can affect investment decisions is with regard to management's perception of outsiders. Surgutneftegaz, for instance, injects 2.5 times more per capita into Surgut city's budget than YUKOS's contribution to the city of Nefteyugansk, where Yuganskneftegaz is located.[25] This may have less to do with high relational capital and secure property rights than the fact that Surgutneftegaz has the most insular, centralized and paternalistic management team among privately owned oil companies in Russia. Its wariness of outsiders is manifested by its being the only major oil company with its corporate headquarters outside of Moscow (it is located in the Siberian town of Surgut). Unlike YUKOS and Sibneft', Surgutneftegaz also eschews employing foreign managers. Decision-making is supercentralized. This insularity and exaggerated sense of paternalism explain Surgutneftegaz's generosity towards Surgut city's social schemes. In contrast, Khodorkovskiy's management philosophy prioritized wages over social benefits, since, in his view, individuals should be left to make their own choices about how to spend the money.[26]

Bogdanov appears to be exceptionally conservative, even among than other *nomenklatura* oilmen, and beholden to the Soviet mentality which perceives that the primary goal of an oil company is to maximize oil production.[27] In contrast, other career oilmen, such as Alekperov and Bogdanchikov, have gradually adopted most of the key tenets of the pro-investor business model – including the maintenance of more transparent accounts, the use of acquisitions to spur growth and openness to foreign investors and partners – as a better way to maximize profits.[28] In other words, it is arguable that the behaviour of 'red directors' is rationally structured by incentives embedded in a specific business model, rather than a product of Soviet-era values, networks and personal relationships.

Rational Choice Institutionalism as an Explanatory Framework

The approach known as rational choice institutionalism has been the main analytical framework used in this book to explain the motivation, calculation and behaviour of, as well as interaction among, individual actors. As noted in the opening chapter, the aim here is not to empirically validate or test this approach. There is no suggestion that rational choice institutionalism provides the final, sole or definitive explanation of political behaviour or policy outcome. Nevertheless it has served as a useful framework to analyse the decision-making process with respect to oil privatization in Russia, and a few examples will be cited to illustrate this point. There are limits to the utility of the approach, but these should not detract from the fact that it is generally suitable for our purposes.

The use of rational choice institutionalism in this book has facilitated explanations of why institutions related to oil privatization were created, perpetuated and sometimes eliminated. The short answer is that self-interested actors established new organizations and rules of conduct in order to realize the gains from cooperation. This is because an institution persists if 'it provides more benefits to the relevant actors than alternate institutional forms',[29] but becomes obsolete if more efficient alternatives are available. A case in point was the different fates of former Soviet ministries that were turned into industry associations, such as Rosneftegaz (oil) and Gazprom (gas). Rosneftegaz was created because the Soviet Oil Ministry did not want to lose its prerogative over the sector to the newly created Russian Mintopenergo. The oil production, refining and marketing entities agreed to join Rosneftegaz only because it had no real power over them; it was not a powerful business *kontsern* as Gazprom, and membership did not preclude these entities from pursuing their interests to merge with other entities to create VIOCs. Rosneftegaz was dissolved just over a year after its founding because Rosneft', unburdened by a Soviet heritage, was perceived as an entity which could better respond to the task of restructuring the oil industry.

In comparison, Gazprom has persisted as an institution because of its continuing utility to its stakeholders. It was and still is a 'cash cow' for its top managers and their relatives. For the state, Gazprom is indispensable to the smooth functioning of the economy, thanks to the implicit subsidies it provides from barter and payment arrears.[30] The company is also a valuable tool to retain and expand Russia's interests abroad: even Nemtsov, one of the most ardent opponents of Gazprom and other natural monopolies, acknowledged that it was 'necessary to preserve Russia's major gas supply company as single entity, since Gazprom . . . is also Russia's visiting card abroad'.[31] As noted in the previous chapter, increasing the state's stake in Gazprom from 38.37% to over 50% was also indicative of Gazprom's value to Putin as a key lever of the economy and of foreign policy and his attempt to use it to guarantee his political longevity by exercising control over it.

Although individuals can craft institutions to suit their interests, as the above discussion indicates, institutions can also structure and direct the behaviour of actors. A case in point concerned decree 1403 on the creation and corporatization of VIOCs in November 1992. One of its provisions was that the state would retain its stakes in the VIOCs for no longer than three years, after which they would be divested in auctions. For the general managers of these newly corporatized companies, it essentially meant that their positions and ownership rights were guaranteed for only three years. In view of uncertainty in the future, it was only rational for them to take immediate advantage of their short-term control over assets and divert them for personal gain.

Another provision of decree 1403 was the creation of a two-tier system within each VIOC, which created holding companies without majority control over their own subsidiaries. This institutional constraint on effective control obliged Muravlenko to engage in a struggle with the heads of YUKOS's subsidiaries, while Khodorkovskiy took four years to buy out minority shareholders within the subsidiaries. This same process was repeated in SIDANKO, TNK and Sibneft', an indication of how the rules of the game, as defined by decree 1403, determined the behaviour of actors in the oil industry.

Or consider the case of BP's merger with TNK in February 2004. BP had clashed with TNK during 1998–1999 over the latter's seizure of key assets belonging to SIDANKO, in which BP had a 10% stake. The fact that BP's negotiations were conducted with the same man (Simon Kukes) who engineered the SIDANKO debacle did not bother BP. As its vice-president explained, the lesson of SIDANKO was that BP 'would design the next deal in Russia in a different way' so as to induce desirable, cooperative, behaviour from its partners.[32]

Nevertheless the main shortcoming of the rational choice institutionalist approach is the assumption 'that institutions are being formed on a *tabula rasa*', to the extent that the 'past history of the institution or the organization is of little concern and a new set of incentives can produce changed behaviours rather easily'.[33] The book has been cognizant of this limitation and where necessary, it supplemented the rational choice institutionalist approach with explanations that took into account an institution's history. This is because an institution's legacy clearly matters for its continued existence. In the case of Rosneftegaz, the fact that much of the personnel were officials of the much-criticized Soviet Oil Ministry reduced its chances of playing a role in the post-Soviet system. By comparison, the fact that a well-regarded expert, instead of the Minister of Coal, was appointed as the head of Rosugol' may have been one factor which contributed to the relative institutional persistence of Rosugol' until 1997.[34] In the case of Rosneft', Putin managed to draw a ring fence around continued state-ownership of Rosneft' as early as September 2000 because of the company's long association as a *de facto* national oil company. However, he did not manage to stake out a similar policy position with regard to ONAKO and Slavneft', which were eventually privatized.

Institutional legacies also impact upon an actor's decision-making calculus and thus cannot be ignored, contrary to the assumptions of the rational choice institutionalist approach. As previously explained, 'young reformers' chose to reform and privatize the oil sector more extensively than the gas sector partly because of different sectoral legacies. For example, unlike the gas industry which was historically managed as a single, unified structure, decision-making in the oil sector was decentralized to four different Soviet ministries. Also, the oil sector was perceived as more open to reform given its history of production problems and attempts to overcome them. Putin's decision to appoint Gutseriev, the head of the BIN group, as president of Slavneft' may also be partly explained by the company's dominance in the political and economic affairs of Ingushetia since 1993: in this regard, Putin hoped to persuade the Ingush to remain neutral in his invasion of Chechnya in 1999 by appointing a leading Ingush to head a major oil company.

Although institutional legacies do matter, they do not by themselves predetermine any outcome. This is because the extent of their influence on decision-making depends on how they are perceived (positively or negatively) by a particular actor. Consider how contrasting appraisals of the state resulted in different policy outcomes. For the 'young reformers', the state was a 'grabbing hand' institution, characterized by unrestrained predation among officials to seize control rights over firms and economic activities. Hence, 'the goal of reform was not to get rid of the managers, but of the ministries'.[35] To this extent, the greatest benefit of corporatization was that it began 'the process of separation of enterprises from the state . . . The enterprise [became] an independent entity, rather than an integral part of some government agency'.[36]

For Putin, this same state is perceived positively. Accordingly, the 'state and its institutions and structures have always played an exceptionally important role in the life of the country and its people. For Russians, a strong state is not an anomaly to be gotten rid of. Quite the contrary, it is a source of order and main driving force of any change'.[37] Putin's reforms have therefore focused on increasing state control over the media, business and economy and on enhancing Moscow's authority over the regions. In other words, institutional legacies are subjective and, therefore, vulnerable to being used as focal points in policymaking to achieve the self-interested ends of relevant actors.

Appendix 1

State's Share in Oil Production, 1994–2006 (as % of Total Oil Production)

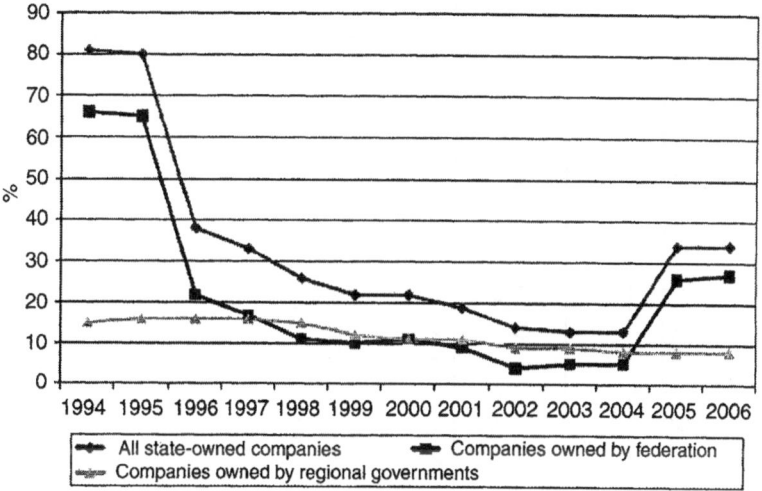

Source: Julia Kuzsnir and Heiko Pleines, 'The Russian Oil Industry between Foreign Investment and Domestic Interests' in *Russian Analytical Digest*, 18 September 2007, p. 14.

Appendix 2
Results of Loans-for-Shares Privatization in the Oil Industry

Name of oil company	Trust auctions, 1995					Collateral auctions, 1996–1997			
	Stake under auction (%)	Loan provided (US$m)	Auction date	Organizer for auction	Auction winner	Min. sale price (US$m)	Auction date	Price paid	Auction winner
Surgut-neftegaz	40.12	88.3	Nov 95	ONEKSIM bank	Surgutneftegaz pension fund	74	Feb 97	78.8	Surgutfondinvest (linked to Surgutneftegaz)
LUKoil	5	35.01	Dec 95	Imperial Bank (LUKoil affiliate)	LUKoil Imperial Bank	43	Jun 97	43.6	LUKoil Reserve-Invest
YUKOS	45	159	Dec 95	Menatep Bank	Laguna (Menatep affiliate)	160	Nov 96	160.1	Monblan (Menatep affiliate)
SIDANKO	51	130	Dec 95	ONEKSIM bank	MFK (part of ONEKSIM bank)	129	Jan 97	129.8	Interros Oil (part of ONEKSIM bank group)
Sibneft'	51	100.1	Dec 95	Menatep Bank	NFK, SBS (linked to Sibneft')	101	May 97	110	FNK (offshoot of NFK)

Sources: Valery Kryukov and Arild Moe. *The Changing Role of Banks in the Russian Oil Sector.* London: Royal Institute of International Affairs, 1998; Nat Moser and Peter Oppenheimer. 'The Oil Industry: Structural Transformation and Corporate Governance'. In *Russia's Post-Communist Economy*, eds, Brigitte Granville and Peter Oppenheimer. Oxford: Oxford University Press, 2001; and *Russian Economic Trends*, 1997.

Notes

1 Introduction to the Policymaking Process

1. During the last few years of the Soviet Union, *kontserny* or commercially managed state enterprises were formed, particularly from departments of ministries.
2. The data in this section are from 'PIW's Top 50: How the Firms Stack Up', *Petroleum Intelligence Weekly* 2003 (available at http://www.energyintel.com); *OECD Economic Surveys 2004: Russian Federation* (Paris: Organisation for Economic Co-operation and Development, 2004), pp. 85–86; and *Russia Oil & Gas Yearbook 2004: Counting Barrels* (Moscow: Renaissance Capital, July 2004), pp. 19, 34 and 81. The remaining volume of oil production and reserves is accounted for by over 100 independent oil producers and joint ventures.
3. See Appendix 1 for the state's changing share in oil production.
4. See Marshall I. Goldman, *The Enigma of Soviet Petroleum: Half-Full or Half-Empty?* (London: George Allen & Unwin, 1980), p. 21. For a brief period, Tsarist Russia even outranked the United States as the world's largest oil producer (ibid., Table 2.1, pp. 14–15). For accounts of the development of the Tsarist and Soviet oil industry by senior industry leaders, see Lev Tchurilov, *Lifeblood of Empire: A Personal History of the Rise and Fall of the Soviet Oil Industry* (New York: PIW Publications, 1996); and Rasul Gouliev, *Oil and Politics: New Relationships among the Oil Producing States – Azerbaijan, Russia, Kazakhstan, and the West* (New York: Liberty Publishing House, 1997).
5. This was a typical characterization of the Soviet economy, as reflected in works by Goldman, *The Enigma of Soviet Petroleum*, p. 91; and Margaret Chadwick, Machiko Nissanke and David Long, *Soviet Oil Exports: Trade Adjustments, Refining Constraints and Market Behaviour* (Oxford: Oxford University Press, 1987), p. 81.
6. Data on oil and gas from Jennifer I. Considine and William A. Kerr, *The Russian Oil Economy* (Cheltenham: Edward Elgar, 2002), Table 5.18, p. 138; and Chadwick, Nissanke and Long, *Soviet Oil Exports*, Table 5.9, pp. 82–83. It appears that Soviet oil exports to the West were deployed as a trade-balancing mechanism to finance imports of grain and technology from those countries, rather than as a foreign policy tool. See ibid., pp. 67–97; and Thane Gustafson, *Crisis amid Plenty: The Politics of Soviet Energy under Brezhnev and Gorbachev* (Princeton, New Jersey: Princeton University Press, 1989), pp. 263–285.
7. Figures cited by Lyuba Pronina, 'Weapons Sales: The Peace Dividend's Evil Twin', *St Petersburg Times*, 25 September 2001; and Aleksey Nikol'skiy, 'Ne kazhdyy god SSSR polluchal takie summy v valyute', *Vedomosti*, 24 October 2005.
8. Yuriy Shafranik and Valeriy Kryukov, *Zapadno-Sibirskiy Fenomen* (Moscow: Neftegazovaya vertikal'', 2000), p. 4. Shafranik is a former general manager of

the Langepas oil production association (1988–1990), governor of Tyumen' oblast' (1991–1992) and Minister of Fuel and Energy (1993–1996). For a similar perspective, see Yegor Gaidar, *Collapse of an Empire: Lessons for Modern Russia*, trans. Antonina W. Bouis (Washington DC: Brookings Institution Press, 2007), pp. 100–109.

9. Quote by a Soviet official as cited by Alexander Arbatov, Vladimir Feygin and Victor Smirnov, 'Unrelenting Oil Addiction', *Russia in Global Affairs*, no. 2 (April–June 2005, available at http://eng.globalaffairs.ru/printver/914.html).

10. See William Tompson, 'The Price of Everything and the Value of Nothing? Unravelling the Workings of Russia's Virtual Economy', *Economy and Society* 28, no. 2 (May 1999); Douglas B. Reynolds, 'Soviet Economic Decline: Did an Oil Crisis Cause the Transition in the Soviet Union?', *The Journal of Energy and Development* 24, no. 1 (2000): 65–81; Stephen Kotkin, *Armageddon Averted: The Soviet Collapse 1970–2000* (Oxford: Oxford University Press, 2001); and Yegor Gaidar, 'The Collapse of the Soviet Union: Lessons for Contemporary Russia', Lecture delivered at the American Enterprise Institute in Washington DC, 13 November 2006.

11. From Goohoon Kwon, 'Post-crisis Revenue Developments in Russia: From an Oil Perspective', *Public Finance and Management* 3, no. 4 (2003), p. 508; *Oil Sector Report* (Moscow: Troika Dialog Research, May 2001), p. 88; and *Russian Federation: Statistical Appendix* (Washington DC: International Monetary Fund, May 2003), Table 31, p. 37.

12. For an explanation of how the data on oil's share in Russia's GDP growth were derived, and why it differs from the 9% cited in official Russian statistics, see *Russian Economic Report* (Moscow: World Bank Group, Russian Federation Country Office, February 2004). Estimates of the oil price elasticity of Russian GDP growth vary widely. The figure cited in the text is used by the World Bank, the basis of which is explained in *From Transition to Development: A Country Economic Memorandum for the Russian Federation* (Moscow: World Bank Group, Russian Federation Country Office, 2004), pp. 10–11. Russia's Ministry of Finance calculates that the effect is less than half of the World Bank's figures, whereas a report by the Bank of Finland claims that it is three times higher. See *OECD Economic Surveys 2004: Russian Federation*, p. 29; and Jouko Rautava, 'The Role of Oil Prices and the Real Exchange Rate in Russia's Economy', in *BOFIT Discussion Papers* (Helsinki: Bank of Finland's Institute for Economies in Transition, 2002), p. 18.

13. Figures on weapons sales are from Konstantin Makienko, 'Financial Results of Russian Arms Trade with Foreign States in 2004', *Moscow Defence Brief*, no. 1 (2005).

14. Lilia Shevtsova, 'The Limits of Bureaucratic Authoritarianism', *Journal of Democracy* 15, no. 3 (2004), pp. 71–72.

15. Today, 90% of coal in Russia is produced by privately owned companies, up from 5% in 1993. See Youri Bobylev and Jacek Cukrowski, 'Russia: Bank Assistance for the Energy Sector', in *World Bank Operations Evaluation Department Working Paper* (Washington DC: World Bank, 2002), p. 23; and Igor Artemiev and Michael Haney, 'The Privatisation of the Russian Coal Industry: Policies and Processes in the Transformation of a Major Industry', in *World Bank Transition Economies Working Paper* (April 2002), p. 6.

16. In June 2005, the state agreed to increase its shareholding in Gazprom to just over 50% at a cost of US$10 billion. The transaction was carried out in stages and was finally settled in December 2005.
17. See Matthew J. Sagers, 'The Russian Natural Gas Industry in the Mid-1990s', *Post-Soviet Geography and Economics* 36, no. 9 (1995), p. 524.
18. See 'PIW's Top 50'.
19. A profile of leading oil producing countries is available from the Energy Information Administration (http://www.eia.doe.gov).
20. S. P. Peregudov, N. Yu Lapina and I. S. Semenenko, *Gruppy interesov i Rossiyskoe gosudarstvo* (Moscow: Editorial URSS, 1999), p. 100.
21. Quote by Putin as cited in Bulat Stolyarov, 'Kreml' khochet zabrat' nedra', *Vedomosti*, 26 July 2002.
22. Philip Hanson and Elizabeth Teague, 'Big Business and the State in Russia', *Europe–Asia Studies* 57, no. 5 (July 2005), p. 662. For a literature survey of the overwhelmingly negative perception of the role of tycoons in Russia's economic development, see Stephen Fortescue, *Russia's Oil Barons and Metal Magnates: Oligarchs and the State in Transition* (Basingstoke: Palgrave Macmillan, 2007), pp. 9–18. The term 'oligarch' (or rule by a few) is widely used in the popular press to refer to Russia's richest businessmen. While the term may have been briefly applicable during the mid-1990s, its continued use in scholarly literature is less appropriate. For this reason, 'tycoons', 'magnates' or 'leading businessmen' are used instead in this book, except where the term 'oligarch' is part of a quotation.
23. A sample of works that focus on the role of big business includes David E. Hoffman, *The Oligarchs: Wealth and Power in the New Russia* (Oxford: Public Affairs, 2002); Chrystia Freeland, *Sale of the Century: The Inside Story of the Second Russian Revolution* (London: Little, Brown and Company, 2000); Juliet Johnson, *A Fistful of Rubles: The Rise and Fall of the Russian Banking System* (Ithaca: Cornell University Press, 2000); Marshall I. Goldman, *The Piratisation of Russia: Russian Reform Goes Awry* (London: Routledge, 2003); Peregudov, Lapina and Semenenko, *Gruppy interesov i Rossiyskoe gosudarstvo*; and Aleksey Mukhin, *Biznes-elita i gosudarstvennaya vlast':Kto vladeet Rossiey na rubezhe vekov?* (Moscow: Centr Politicheskoy Informatsii, 2001).
24. For empirical studies, see Daniel Bollinger, 'The Four Cornerstones and Three Pillars in the "House of Russia" Management System', *Journal of Management Development* 13, no. 2 (1994): 49–54; Alexander I. Naumov, 'Hofstede's Measurement of Russia: The Influence of National Culture on Business Management', *Management* 1, no. 3 (1996): 70–103; Detelin S. Elenkov, 'Differences and Similarities in Managerial Values between US and Russian Managers: An Empirical Study', *International Studies of Management and Organisation* 27, no. 1 (1997): 85–106; and Fred Luthans, Dianne H. B. Welsch and Stuart A. Rosenkratz, 'What do Russian Managers Really Do? An Observational Study with Comparisons to US Managers', *Journal of International Business Studies* 24, no. 4 (1993), p. 742. For a discussion of the differences between *nomenklatura* managers and Western managers, see Geert Hofstede, 'Cultural Constraints in Management Theories', *The Academy of Management Executive* 7, no. 1 (1993): 81–94; and Sheila M. Puffer, 'Shedding the Legacy of the Red Executive: Leadership in Russian Enterprises', *International Business Review* 4, no. 2 (1995): 157–176.

25. David Holt, David A. Ralston and Robert H. Terpstra, 'Constraints on Capitalism in Russia: The Managerial Psyche, Social Infrastructure, and Ideology', *California Management Review* 36, no. 3 (1994), p. 129.

26. In comparison, there is a large volume of works on the role of energy companies in Russian foreign policy. They are variously perceived as agents of a neoimperialist foreign policy, or of using the state to further their commercial interests within the Commonwealth of Independent States and the West, or of acting in partnership with the state to forge a foreign policy that combines global, national and corporate interests. See, for example, Amy Meyers Jaffe and Robert A. Manning, 'Russia, Energy and the West', *Survival* 43, no. 2 (Summer 2001): 133–152; Douglas W. Blum, 'Domestic Politics and Russia's Caspian Policy', *Post-Soviet Affairs* 14, no. 2 (1998): 137–164; Robert Stowe, 'Foreign Policy Preferences of the New Russian Business Elite', *Problems of Post-Communism* 48, no. 3 (2001): 49–58; and Yakov Pappe, 'Neftyanaya i gazovaya diplomatiya Rossii', *Pro et Contra*, no. 3 (1997), available at http://www.carnegie.ru/ru/pubs/procontra/55622.htm.

27. See Eugene Khartukov, 'Russia's Oil Majors: Engine for Radical Change', *Oil and Gas Journal*, 27 May 2002; Eugene Khartukov, 'Russia's Oil Privatisation is more Greed than Fear', *Oil and Gas Journal*, 3 July 2000; Nat Moser and Peter Oppenheimer, 'The Oil Industry: Structural Transformation and Corporate Governance', in *Russia's Post-Communist Economy*, ed. Brigitte Granville and Peter Oppenheimer (Oxford: Oxford University Press, 2001); Valery Kryukov and Arild Moe, *The Changing Role of Banks in the Russian Oil Sector* (London: Royal Institute of International Affairs, 1998); David Lane, 'The Russian Oil Elite: Background and Outlook', in *The Political Economy of Russian Oil*, ed. David Lane (Lanham: Rowman and Littlefield, 1999); and Sergei P. Peregudov, 'Large Corporations as National and Global Players: The Case of Lukoil', in *Explaining Post-Soviet Patchworks: Actors and Sectors in Russia between Accommodation and Resistance to Globalisation (volume 1)*, ed. Klaus Segbers (Aldershot: Ashgate, 2001).

28. Charles E. Lindblom, *Politics and Markets: The World's Political–Economic Systems* (New York: Basic Books, 1977), p. 194. See also Ralph Miliband, *The State in Capitalist Society* (London: Weidenfeld & Nicholson, 1969).

29. Donald D. Jensen, 'How Russia is Ruled', in *Business and the State in Contemporary Russia*, ed. Peter Rutland (Boulder: Westview Press, 2001), p. 53.

30. Ibid., p. 53.

31. David Hoffman E., 'Putin Aims to Assure Tycoons: Russian Disavows Pressure by Police', *The Washington Post*, 29 July 2000.

32. Lindblom, *Politics and Markets*, p. 172.

33. David Fairlamb, 'Reining in the Oligarchs: Russia', *International Investor*, 1 November 1998.

34. Author's interview with Mr Nick Butler, group vice-president (Strategy), British Petroleum, 20 January 2005, London. The TNK–BP merger is Russia's largest foreign investment in the country to date.

35. As estimated by Peter Boone and Denis Rodionov, 'Rent Seeking in Russia and the CIS' (paper presented at the Tenth Anniversary Conference of the European Bank for Reconstruction and Development, London, December 2001), p. 3.

36. Chapter 3 contains a further discussion of the vulnerability of the Duma to inducements from other actors.

37. The figures are cited by Hoffman, *The Oligarchs*, p. 350.

38. Hoffman, *The Oligarchs*, p. 351.

39. For Khodorkovskiy's pre-1992 profile, see Hoffman, *The Oligarchs*, pp. 105–119; and for Potanin's see Freeland, *Sale of the Century*, pp. 121–126. Tycoons who were not part of the Soviet *nomenklatura* include Boris Berezovskiy (oil), Vladimir Gusinskiy (media magnate) and Aleksandr Smolenskiy (banker). For an account of how they became leading businessmen, see Hoffman, *The Oligarchs*.

40. See 'LUKoil officials win key Government Oil Posts', *Petroleum Economist*, July 2000.

41. 'Insiders' are groups that consider it important to develop close relations with government officials, and that have gained acceptance and been granted subsequent access by these officials. See discussion by Wyn Grant, 'The Role and Power of Pressure Groups', in *British Politics in Perspective*, ed. R. L. Borthwick and J. E. Spence (Leicester: Leicester University Press, 1984), pp. 132–138; and Jordan and Richardson, *Government and Pressure Groups*, pp. 30–31. For a discussion of the role of RSPP since 2000, see Tina Podplatnik, 'Big Business and the State in Putin's Russia, 2000–2004: Towards a New State Corporatism' (D.Phil. thesis, University of Oxford, 2005); Hanson and Teague, 'Big Business and the State in Russia', pp. 658–665; and Aleksey Zudin, 'Neo-koporativizm v rossiya? Gosudarstvo i biznes pri Vladimire Putine', *Pro et Contra* 6, no. 4 (Fall 2001): 171–198.

42. See the discussions by Mancur Olson, *The Logic of Collective Action: Public Goods and the Theory of Groups* (Cambridge, Massachusetts: Harvard University Press, 1971); and Mancur Olson, *The Rise and Decline of Nations: Economic Growth, Stagflation and Social Rigidities* (New Haven: Yale University Press, 1982).

43. Quote by Leonid Nevzlin, a partner in Menatep, a holding company which acquired YUKOS as a result of the auctions, as cited in Freeland, *Sale of the Century*, p. 166.

44. Olson, *The Rise and Decline of Nations*, p. 62. The same argument as applied to the post-communist transition is contained in Mancur Olson, *Power and Prosperity: Outgrowing Communist and Capitalist Dictatorships* (New York: Basic Books, 2000).

45. See Joel S. Hellman, 'Winners Take All: The Politics of Partial Reform in Postcommunist Transitions', *World Politics* 50, no. 2 (1998): 203–234. Similar arguments are advanced by Olson, *Power and Prosperity*; Anders Aslund, 'Why Has Russia's Economic Transformation Been So Arduous?' (paper presented at the Annual World Bank Conference of Development Economics, Washington DC, 28–30 April 1999); and Robinson, 'The myth of equilibrium'.

46. Boone and Rodionov, 'Rent Seeking in Russia and the CIS', p. 20.

47. Quote by Andrey Melnichenko, Chairman of MDM Bank, cited in 'Russia: Back from the Brink', *Institutional Investor*, 1 February 2002.

48. As noted by David Vogel, *Fluctuating Fortunes: The Political Power of Business in America* (New York: Basic Books, 1989), p. 7; and Edwin M. Epstein, *The Corporation in American Politics* (Englewood Cliffs, New Jersey: Prentice-Hall, 1969), pp. 192, 221–222.

49. Graham K. Wilson, *Business and Politics: A Comparative Introduction* (Basingstoke: Palgrave Macmillan, 1990), p. 183.
50. Andrei Shleifer and Maxim Boyko, 'The Politics of Russian Privatisation', in *Post-Communist Reform: Pain and Progress*, ed. Oliver Blanchard et al. (Cambridge, Massachusetts: MIT Press, 1993), p. 39.
51. Andrei Shleifer and Daniel Treisman, *Without a Map: Political Tactics and Economic Reform in Russia* (Cambridge, Massachusetts: MIT Press, 2000), p. 17.
52. Quote by Leonid Abalkin, economic adviser to Nikolay Ryzhkov in the Gorbachev era, as cited in Lynn Nelson and Irina Kuzes, *Radical Reform in Yeltsin's Russia: Political, Economic and Social Dimensions* (New York: M E Sharpe, 1995), pp. 135–137.
53. A further discussion of this issue is in Chapter 2.
54. See Konstantin Smirnov, 'Staynyy instinkt', *Kommersant'-Daily*, 27 August 2001; and Zudin, 'Neo-koporativizm v rossiya?'.
55. For an in-depth discussion of bureaucratic rivalries as a form of 'bureaucratic capitalism', see Nodari Simonia, 'Economic Interests and Political Power in Post-Soviet Russia', in *Contemporary Russian Politics: A Reader*, ed. Archie Brown (Oxford: Oxford University Press, 2001).
56. Sergei Peregudov and Irina Semenenko, 'Lobbying Business Interests in Russia', *Democratisation*, 3, no. 2 (1996), p. 133.
57. Shleifer and Treisman, *Without a Map*, p. 17.
58. Ibid., p. 17.
59. Georgiy Bovt, Sergey Zhikharev and Natal'ya Kalashnikova, 'Vlasti ne stesnyayutsya – za nee boryutsya', *Kommersant'-Daily*, 16 August 1995.
60. Stephen Sestanovich, 'Force, Money, and Pluralism', *Journal of Democracy* 15, no. 3 (2004), p. 42.
61. The book will adopt the term 'rational choice institutionalism' as favoured by B. Guy Peters, *Institutional Theory in Political Science: The 'New Institutionalism'* (London: Pinter, 1999); and Peter A. Hall and Rosemary C. R. Taylor, 'Political Science and the Three New Institutionalisms', *Political Studies* XLIV, no. 5 (1996): 936–957. Other terms to describe the same analytical framework include 'actor-centred institutionalism' (Scharpf), 'structure-induced equilibrium' (Shepsle) and 'actor-system dynamics' (Burns et al.). See Fritz W. Scharpf, *Games Real Actors Play: Actor-Centred Institutionalism in Policy Research* (Boulder: Westview Press, 1997); Kenneth A. Shepsle, 'Studying Institutions: Some Lessons from the Rational Choice Approach', *Journal of Theoretical Politics* 11, no. 2 (1989): 131–147; and Tom Burns, Thomas Baumgartner and Philippe Deville, *Man, Society, Decisions: The Theory of Actor-System Dynamics for Social Scientists* (New York: Gordon and Breach, 1985).
62. See works by Stefan Hedlund, *Russia's 'Market' Economy: A Bad Case of Predatory Capitalism* (London: University College of London Press, 1999); Michael McFaul, Nikolai Petrov and Andrei Ryabov, 'Introduction', in *Between Dictatorship and Democracy: Russian Post-Communist Political Reform* (Washington DC: Carnegie Endowment for International Peace, 2004); Michael McFaul, 'Institutional Design, Uncertainty and Path Dependency during Transitions: Cases from Russia', *Constitutional Political Economy* 10, no. 1 (1999): 27–52; Timothy Frye, 'A Politics of Institutional Choice: Post-Communist Presidencies', *Comparative Political Studies* 30, no. 5 (1997): 523–552; Maxim Boyko, Andrei Shleifer and Robert W. Vishny, *Privatising*

Russia (Cambridge, Massachusetts: MIT Press, 1995); and Shleifer and Treisman, *Without a Map*; and Yuko Adachi, 'Reconstitution of Post-Soviet Ex-State Enterprises into Russian Business Firms under Institutional Weakness', in *Centre for the Study of Economic and Social Change in Europe Working Paper no. 56* (School of Slavonic & East European Studies, London: July 2005).

63. For further discussion on the role of individual actors in politics, and how they have shaped institutions, see Keith Dowding, 'The Compatability of Behaviouralism, Rational Choice and "New Institutionalism"', *Journal of Theoretical Politics* 6, no. 1 (1994): 105–117; Steven Lee Solnick, *Stealing the State: Control and Collapse in Soviet Institutions* (Cambridge, Massachusetts: MIT Press, 1998); Juan Linz and Alfred Stepan, *Problems of Democratic Transition and Consolidation: Southern Europe, South America and Post-Communist Europe* (Baltimore: The Johns Hopkins University Press, 1996); and Michael Burton, Richard Gunther and John Higley, 'Elites and Democratic Consolidation in Latin America and Southern Europe: An Overview' in John Higley and Richard Gunther (eds), *Elites and Democratic Consolidation in Latin America and Southern Europe*, Cambridge: Cambridge University Press, 1992.

64. William Roberts Clark, 'Agents and Structures: Two Views of Preferences, Two Views of Institutions', *International Studies Quarterly* 42, no. 2 (1998), p. 249.

65. John C. Harsanyi, 'Rational-Choice Models of Political Behaviour vs. Functionalist and Conformist theories', *World Politics* 21, no. 4 (1969), p. 518. For a similar perspective, see Mark Irving Lichbach, *Is Rational Choice Theory All of Social Science?* (Ann Arbor: University of Michigan Press, 2003), p. 59.

66. Herbert A. Simon, 'A Behavioural Model of Rational Choice', *The Quarterly Journal of Economics*, LXIX (1955), p. 118. The word 'satisficing' is a combination of 'satisfied' and 'sufficient'.

67. Scharpf, *Games Real Actors Play*, p. 42. The definition of institutions as formal and informal structures is as per Larry L. Kiser and Elinor Ostrom, 'The Three Worlds of Actions: A Metatheoretical Synthesis of Institutional Approaches', in *Strategies of Political Inquiry*, ed. Elinor Ostrom (Beverley Hills, California: SAGE Publications, 1982); and Douglass C. North, 'The Contribution of the New Institutional Economics to an Understanding of the Transition Problem' (paper presented at the WIDER Annual Lectures 1, Helsinki, March 1997).

68. Barnes relies more on an institutionalist-centred framework to explain the redistribution of property in the agriculture and industrial sectors, noting that the motives and strategies of participants were shaped by the 'incentives and opportunities the environment provides'. See Andrew Barnes, *Owning Russia: The Struggle over Factories, Farms and Power* (Ithaca: Cornell University Press, 2006), p. 232. While a rational choice institutionalist approach does not dispute this relationship, the difference is that it allows as well for the possibility that an individual can reshape the environment to his own benefit.

69. For criticisms of rational choice institutionalism as an approach, see Donald Green and Ian Shapiro, *Pathologies of Rational Choice Theory: A Critique of Applications in Political Science* (New Haven: Yale University Press, 1994).

Responses to the critique and a rejoinder are contained in Jeffrey Friedman (ed.), *The Rational Choice Controversy: Economic Models of Politics Reconsidered* (New Haven: Yale University Press, 1995).

70. See Scharpf, *Games Real Actors Play*; and Peter A. Hall and David Soskice, 'An Introduction to Varieties of Capitalism', in *Varieties of Capitalism: The Institutional Foundations of Comparative Advantage*, ed. Peter A. Hall and David Soskice (Oxford: Oxford University Press, 2001), pp. 6–7. A further discussion of the concept of a composite actor is found in James S. Coleman, *Individual Interests and Collective Action* (Cambridge: Cambridge University Press, 1986); and James S. Coleman, *Power and Structure of Society* (New York: W W Norton & Company, 1974).

71. As cited by Valery Kryukov and Arild Moe, *Gazprom: Internal Structure, Management Principles and Financial Flows* (London: Royal Institute of International Affairs, 1996), p. 24.

72. Although gas may also be transported in liquefied form, the volume of this trade at present is very low given the huge capital investments required and is imported mostly by Japan.

73. For a location of the gas fields of Urengoy, Yamburg and Nadym, see the map in *Energy Policies of the Russian Federation*, p. 166.

74. Ibid., p. 101. They included Surgutneftegaz, Yuganskneftegaz and Nizhnevartovsk.

75. See, for example, Dmitriy Dokuchaev, 'Gazprom Reform: Hard Nut to Crack', *Moscow News*, 16 July 2003. Chernomyrdin likewise noted that 'nothing good would have resulted from such a division' of Gazprom into production and transportation companies as proposed by GKI (from author's interview with Mr Viktor Chernomyrdin).

76. See Gary S. Becker, 'There's Nothing Natural about "Natural" Monopolies', *Business Week*, 6 October 1997, p. 26; and S. Ran Kim and A. Horn, 'Regulating Policies concerning Natural Monopolies in Developing and Transition Economies', in *DESA Discussion Paper no. 8* (March 1999).

77. The key actors in the Soviet oil industry comprised the various oil-related ministries – the Ministry of Oil Industry, Ministry for Construction of Oil and Gas Enterprises, Ministry of Geology and the State Planning Ministry – along with several individuals known as 'oil generals'. In contrast, there was only one actor of consequence in the Soviet gas sector, that is, the Ministry of Gas and later, the Chairman of the Gazprom *kontsern*.

78. For examples of interministerial conflicts about the Soviet oil industry, see Gustafson, *Crisis amid Plenty*, pp. 291–301.

79. Calculated from *A Study of the Soviet Economy: Volume 3* (Washington DC: International Monetary Fund, 1991), Table V6.4, p. 226.

80. See *Russia Energy Survey*, Table 3.1, p. 49.

81. These figures are calculated based on the reserve-to-production ratios at the end of 2003. Proved oil reserves stood at 69 billion barrels whereas proved gas reserves amounted to 277 billion barrels of oil equivalent. See *Russian Oil & Gas Yearbook 2004*, pp. 9–10.

82. From Andrei Konoplyanik and Nikolai Lisovsky, 'Russia Aims for Favourable Climate for Joint Ventures', *Oil & Gas Journal*, 10 August 1992. The classic statement on the role of revenue in the making of state policy is by Margaret Levi, *Of Rule and Revenue* (Berkeley: University of California Press, 1988).

For an analysis of the role of revenue as applied to privatization strategies in post-communist transitions, see Gerald M. Easter, 'Politics of Revenue Extraction in Post-Communist States: Poland and Russia Compared', *Politics & Society* 30, no. 4 (2002): 599–627; and Pauline Jones Luong and Erika Weinthal, 'Contra Coercion: Russian Tax Reform, Exogenous Shocks, and Negotiated Institutional Change', *American Political Science Review* 98, no. 1 (February 2004): 139–152. On the link between revenue and institutional reform, see Atsushi Ogushi, 'Money, Property and the Demise of the CPSU', *Journal of Communist Studies and Transition Politics*, 21, no. 2 (June 2005): 268–295.

83. Author's interview with Mr Viktor Chernomyrdin, former chairman of Gazprom and ex-prime minister of the Russian Federation, 11 February 2004, Kiev.

84. Decree 1333 'On the transformation of the state gas *kontsern*, Gazprom, into the Russian joint-stock company, Gazprom' was issued on 5 November 1992. It provided for Gazprom to be corporatized as a single and fully intact enterprise. This differed from the situation in the oil industry which was divided into several separate enterprises as part of corporatization.

85. For a comprehensive account of early reforms in the gas industry, see Kryukov and Moe, *Gazprom: Internal Structure, Management Principles and Financial Flows*, pp. 24–41.

86. See, for instance, Vladimir Putin, 'Russia at the Turn of the Millennium', (Moscow: 31 December 1999); and Roy Medvedev, *Post-Soviet Russia: A Journey Through the Yeltsin Era*, trans. George Shriver (New York: Columbia University Press, 2000), p. 54.

87. On the use of case studies in research, see Barbara Geddes, 'How the Cases you Choose Affect the Answers You Get: Selection Bias in Comparative Politics', in *Political Analysis: An Annual Publication of the Methodology Section of the American Political Science Association*, James A. Stimson (ed.) (Ann Arbor: University of Michigan Press, 1990); Arend Lijphart, 'The Comparable Cases Strategy in Comparative Research', *Comparative Political Studies*, no. 8 (1975): 682–693, Arend Lijphart, 'Comparative Politics and Comparative Method', *American Political Science Review* 65 (1971): 158–177; Arthur Stinchcombe, *Theoretical Methods in Social History* (New York: Academic Press, 1978); Adam Przeworski and Henry Tenue, *The Logic of Comparative Social Inquiry* (New York: John Wiley, 1970); and Theda Skocpol, *States and Social Revolutions: A Comparative Analysis of France, Russia and China* (Cambridge: Cambridge University Press, 1979).

2 The Initiation and Spread of Privatization

1. Harvey B. Feigenbaum and Jeffrey R. Henig, 'The Political Underpinnings of Privatisation: A Typology', *World Politics* 46, no. 2 (1994), p. 185.

2. For an enumeration of the variety of privatization activities, see Graeme A. Hodge, *Privatisation: An International Review of Performance* (Boulder: Westview, 2000), Peter M. Jackson and Catherine Price, 'Privatisation and Regulation: A Review of the Issues', in *Privatisation and Regulation: A Review of the Issues*, ed. Peter M. Jackson and Catherine Price (London: Longman, 1994); and 'Types and Techniques of Privatisation', *Privatisation Database* (available at http://www.privatization.org).

3. Similar definitions of privatization are employed by William L. Megginson and Jeffrey M. Netter, 'From State to Market: A Survey of Empirical Studies on Privatisation', *Journal of Economic Literature* 39, no. 2 (June 2001): 321–389; Joseph C. Brada, 'Privatisation is Transition – Or Is It?', *Journal of Economic Perspectives* 10, no. 2 (1996): 67–86; and Sunita K. Kikeri, John Nellis and Mary S. Shirley, *Privatisation: The Lessons of Experience* (Washington DC: World Bank, 1992), p. 14.

4. One of the key debates about the Soviet energy industry concerned the extent to which its production decline was the result of natural depletion or mismanagement. For an overview of the problems in the Soviet energy industry from the perspective of a former Russian Minister of Foreign Economic Relations, see Andrei Konoplyanik, 'Russia Struggling to Revive Production, Rebuild Oil Industry', *Oil & Gas Journal*, 2 August 1993. See also previously cited works by Considine and Kerr, *The Russian Oil Economy*; Ebel, *Energy Choices in Russia*; and Gustafson, *Crisis amid Plenty*.

5. Annual oil production figures are provided in *Energy Policies of the Russian Federation* (Paris: International Energy Agency, 1995), Figure 3, p. 107; and *World Production of Crude Oil, Natural Gas Plant Liquids, Other Liquids and Refinery Processing: 1980–2001* (Energy Information Administration, US Department of Energy, available at http://www.eia.doe.gov).

6. See Valeriy Neverov and Aleksandr Igolkin, 'Neft' rodiny', *Ekonokima i zhizn'*, December 1991.

7. As cited in *Energy Policies of the Russian Federation*, p. 102.

8. Russia's energy intensity use was measured at 1.10 tonnes of oil equivalent (or toe) per US$1000, while the OECD average was 0.23 toe/US$1,000. Canada, which provides a better comparison in view of its harsh climate and vast territory, records an energy intensity of 0.39 toe/US$1000 or one-third the level in Russia. See ibid., pp. 43–44.

9. Calculated from information provided by Anders Aslund, *Building Capitalism: The Transformation of the Former Soviet Bloc* (Cambridge: Cambridge University Press, 2002), Table 1.1, p. 35. To date, there is no consensus on actual Soviet growth rates and recomputations are still ongoing. According to the OECD, for instance, Soviet growth rates during the 1960s, 1970s and 1980s were 5%, 2% and 1.2%, respectively, as cited by Hans Aage, 'Privatisation and Democratisation in Eastern Europe and the Former Soviet Union', in *International Privatisation: Strategies and Practices*, ed. Thomas Clarke (Berlin: Walter de Gruyter, 1994), Table 9.1, p. 167. Notwithstanding the point here is that from the 1970s the Soviet economy continuously declined, in contrast to the growth in the economies of the Western and developing countries.

10. For instance, oil's share of capital investment in Soviet industry amounted to an annual average of 50% between 1981 and 1990, as noted by Considine and Kerr, *The Russian Oil Economy*, Table 6.3, p. 151. The link between energy policy and economic stagnation in the Soviet economy is discussed by Leslie Dienes, 'The Energy System and Economic Imbalances in the USSR', *Soviet Economy* 1, no. 4 (1985): 340–372; Boris Rumer, 'Structural Imbalance in the Soviet Economy', *Problems of Communism* 33 (July–August 1984): 24–32; and Gertrude E. Schroeder, 'The Slowdown in Soviet Industry, 1976–1982', *Soviet Economy* 1, no. 1 (1985): 42–74.

11. For details on these laws, see Simon Johnson and Heidi Kroll, 'Managerial Strategies for Spontaneous Privatisation', *Soviet Economy* 7 (1991): 281–316.

12. Quote by Vagit Alekperov, President of LUKoil, as cited by David Lane, 'The Political Economy of Russian Oil', in *Business and the State in Contemporary Russia*, ed. Peter Rutland (Boulder: Westview Press, 2001), p. 103.

13. For a comprehensive discussion about the role of industrial ministries in the Soviet system, see Stephen Whitefield, *Industrial Power and the Soviet State* (Oxford: Clarendon Press, 1993); and Hough, *The Logic of Economic Reform in Russia*, pp. 32–36.

14. The concept of a *kontsern* approximates that of a commercialized state-owned enterprise. Such enterprises are profit oriented and have more flexibility in operational issues than state-owned ones. However, the company law that dictates the management of all private companies is not legally binding on commercialized enterprises. Commercialization is thus less far-reaching than corporatization, which occurs when an enterprise is converted into a company and bound by the provisions of company law, but is still a property of the state. Commercialization/corporatization may be ends in themselves or an intermediate step towards privatization.

15. Whitefield, *Industrial Power and the Soviet State*, p. 227.

16. Author's interview with Mr Lev Churilov, former Minister of Oil of the USSR, 12 February 2004, Moscow.

17. Moser and Oppenheimer, *The Oil Industry*, p. 304.

18. The official, Soviet-era title, for the heads of enterprises was *general-direktor*. In the book, we will replace this with the more familiar term 'general manager'.

19. Author's interview with Mr Aleksandr Putilov, former President of Rosneft', 3 October 2003, Moscow.

20. From 'O kompanii: Istoriya', in *LUKoil neftyanaya kompaniya* (available on the company's homepage at http://www.lukoil.ru). Alekperov's role in the reorganization of the oil industry is also discussed in 'Glava 'LUKoyla' dal interv'yu: Gosudarstvo otvechaet kompanii vzaimnost'yu', *Kommersant'-Daily*, 6 February 1996.

21. Up to 20% of Russia's oil was being smuggled out of the country in 1992 and as many as 50 railway cars, each with a capacity of 150 barrels of oil, disappeared each day in 1993, according to Svetlana P. Glinkina, 'Privatizatsiya and Kriminalizatsiya: How Organized Crime is Hijacking Privatization', *Democratizatsiya* 2, no. 3 (Summer 1994), p. 389. Oil smuggling still continues till this day, albeit in smaller quantities than previously, as noted by Sergey Karnaukhov, 'Neft' i kriminal', *Neftegazovaya Vertikal'*, (no. 2) 2000. The link between the KGB and organized crime has been the subject of many works, including J. Michael Waller, *Soviet Empire: The KGB in Russia Today* (Boulder: Westview Press, 1994); Stephen Handelman, *Comrade Criminal: Russia's New Mafiya* (Yale: Yale University Press, 1995); and Gregoriy Podlesskikh and Andrey Tereshonok, *Vory v zakone: brosok k vlasti* (Moscow: Khudozhestvennaya Literatura, 1994).

22. The extent of the price differential is taken from Aslund, *Building Capitalism*, p. 172.

23. The estimate is by Vladimir Tikhomorov, 'Capital Flight from Post-Soviet Russia', *Europe–Asia Studies* 49, no. 4 (1997), Table 3, p. 605. One general manager noted that in 1991–1992 'one litre of mineral water was 100 times

more expensive than a litre of oil'. See Anatoliy Sivak, 'Where Will the Heavy Pressure of the Ministry of Fuel and Energy Lead?', *Official Kremlin International News Broadcast*, 22 April 1992.

24. On the relationship between policy change and crises, see Joan M. Nelson, 'Conclusion', in *Economic Crisis and Policy Choice: The Politics of Adjustment in the Third World*, ed. Joan M. Nelson (Princeton: Princeton University Press, 1990), p. 325; Ruth Berins Collier and David Collier, *Shaping the Political Arena: Critical Junctures, the Labour Movement and Regime Dynamics in Latin America* (Princeton: Princeton University Press, 1991); Merilee S. Grindle and John W. Thomas, *Public Choices and Policy Change* (Baltimore: Johns Hopkins University Press, 1991); and Stephen D. Krasner, 'Approaches to the State: Alternative Conceptions and Historical Dynamics', *Comparative Politics* 16, no. 2 (January 1984): 223–246.

25. Tchurilov, *Lifeblood of Empire*, p. 186. The tendency to blame production shortfalls on poor 'work discipline' continued under Andropov and Gorbachev, so that by the mid-1980s, the average tenure of a chief engineer in an oil production entity fell to just over one year. From Gustafson, *Crisis amid Plenty*, p. 103.

26. Discussions that culminated in the 1983 energy programme were initiated under Brezhnev and were probably partly a response to a report by the Central Intelligence Agency of the United States, predicting that oil-production levels in the Soviet Union would decline drastically to the extent that the Soviet Union would have to import oil for its own needs by the mid-1990s. See the declassified report by the Central Intelligence Agency, 'The Impending Soviet Oil Crisis', (1977, available at http://164.109.56.133/soviet-intelligence.asp). Gustafson has argued that while the crisis in the Soviet oil industry was partly the result of problems and biases inherent in the command economy, it was aggravated by the character of Brezhnev as a leader, who chose not to tackle such issues decisively. See Gustafson, *Crisis amid Plenty*.

27. Cited by Shafranik and Kryukov, *Zapadno-Sibirskiy Fenomen*, p. 18.

28. Stephan Haggard and Robert R. Kaufman, 'Introduction: Institutions and Economic Adjustment', in *The Politics of Economic Adjustment: International Constraints, Distributive Conflicts and the State*, ed. Stephan Haggard and Robert R. Kaufman (Princeton: Princeton University Press, 1992), p. 21.

29. On the influence of experts and their ideas in spreading similar reform policies around the world, see Peter M. Haas, 'Introduction: Epistemic Communities and International Policy Coordination', *International Organisation* 46, no. 1 (1992): 1–35; Miles Kahler, 'External Influence, Conditionality and the Politics of Adjustment', in *The Politics of Economic Adjustment: International Constraints, Distributive Conflicts and the State*, ed. Stephan Haggard and Robert R. Kaufman (Princeton: Princeton University Press, 1992); Barbara Stallings, 'International Influence on Economic Policy: Debt, Stabilisation and Structural Reform', in ibid.; Margaret E. Keck and Kathryn Sikkink, *Activists beyond Borders: Advocacy Networks in International Politics* (Ithaca: Cornell University Press, 1998); Peter A. Hall, ed., *The Political Power of Economic Ideas: Keynesianism across Nations* (Princeton: Princeton University Press, 1989); and G. John Ikenberry, 'The International Spread of Privatisation Policies: Inducements, Learning and "Policy Bandwagoning"', in *The Political Economy of Public Sector Reform and Privatisation*, ed. Ezra N. Suleiman and John Waterbury (Boulder: Westview,

1990). For a critique of the work of international agencies in Russia, see Joseph E. Stiglitz, 'Whither Reform? Ten Years of Transition' (paper presented at the Annual World Bank Conference on Development Economics, Washington DC, 28–30 April 1999); and Vladimir Brovkin, 'Wishful Thinking about Russia?', *Transition* (June 1999). Criticism of the IMF and World Bank has also come from liberal economists who claim that these agencies did not do enough to support the 'young reformers' at the beginning of the reforms. See Jeffrey D. Sachs, 'The Transition at Mid Decade', *American Economic Review* 86, no. 2 (1996): 128–133.

30. Prior to Thatcher's privatization programme, west Germany under Konrad Adenauer disposed of governmental shares in two major industrial companies in the 1960s, and the Labour government in the UK sold shares in British Petroleum to raise cash in 1976. However, the privatization programme launched under Thatcher is considered to be the most important historically.

31. John Nellis, 'Privatisation in Developing Countries: A Summary Assessment', in *Centre for Global Development Working Paper 87* (Washington DC: 2006), p. 3.

32. See Megginson and Netter, 'From State to Market: A Survey of Empirical Studies on Privatisation', pp. 327–328.

33. Nellis, 'Privatisation in Developing Countries', p. 8.

34. According to Fariborz Ghadar, the oil privatizations in the 1980s and early 1990s were partly the result of industry-wide factors relating to the increasing indebtedness of national oil companies in the face of falling world demand, and consequently, world price for oil. However, the Middle Eastern governments managed to retain control over their oil industries because of the comparatively lower production costs and levels of accountability in the region. See Fariborz Ghadar, 'Oil: The Power of an Industry', in *The Promise of Privatisation: A Challenge for US Policy*, ed. Raymond Vernon (Washington DC: Council of Foreign Relations, 1988).

35. Calculated from figures provided in *World Production of Crude Oil*.

36. Alexander Radygin, *Privatisation in Russia: Hard Choice, First Results, New Targets* (London: The Centre for Research into Communist Economies, 1995), p. 20. A review of competing reform plans and tendencies is given in ibid., pp. 20–26; Goldman, *Lost Opportunity*, pp. 64–77; Shlapentokh, 'Privatisation Debates in Russia: 1989–1992'; and Vladimir Mau, *The Political History of Economic Reform in Russia, 1985–1994* (London: The Centre for Research into Communist Economies, 1996).

37. See George W. Breslauer, 'Soviet Economic Reforms since Stalin: Ideology, Politics and Learning', *Soviet Economy* 6, no. 3 (1990): 252–280.

38. The estimated proportion of enterprises exempted from the mass privatization programme is cited in Anders Aslund, *How Russia Became a Market Economy* (Washington DC: Brookings Institution Press, 1995), p. 233; Peter Rutland, 'Privatisation in Russia: One Step Forward: Two Steps Back?', *Europe–Asia Studies* 46, no. 7 (1994), p. 1119; and *OECD Economic Surveys: The Russian Federation 1995* (Paris: Organisation for Economic Co-operation and Development, 1995), p. 81.

39. On the role of domestic and foreign actors, see Haggard and Kaufman, 'Introduction: Institutions and Economic Adjustment'; and the list of works cited by Robert H. Bates and Anne O. Krueger, 'Generalisations from the Country Studies', in *Political and Economic Interactions in Economic Policy*

Reform: Evidence from Eight Countries, ed. Robert H. Bates and Anne O. Krueger (Oxford: Blackwell, 1993), footnote 6, p. 468.

40. Vladimir Mau, 'Russian Economic Reforms as Perceived by Western Critics', in *BOFIT Discussion Papers* (Helsinki: Bank of Finland's Institute for Economies in Transition, December 1999), p. 15. For a similar perspective, see John Odling-Smee, 'The IMF and Russia in the 1990s', in *IMF Working Paper* (Washington DC: August 2004).

41. As cited by Igor Ognyov, 'Tyumen's Surprise for Russia', *Izvestiya*, 9 January 1992 (compiled in *Current Digest of the Post-Soviet Press* XLIV, 1, 1992, pp. 25–26, 35).

42. Aspects of the Gaydar/Chubays plan for the oil industry are cited in Peregudov, 'Large Corporations as National and Global Players', p. 122; Lane, 'The Political Economy of Russian Oil', p. 106; and in author's interview with Mr Dmitriy Romanov.

43. This disagreement between Gaydar/Chubays and Lopukhin was not the only instance of differences among the 'young reformers'. Gaydar and Chubays, for example, opposed the use of vouchers for mass privatization, a policy championed by Dmitriy Vasil'ev, the deputy head of GKI. They were instead, in favour of cash-based privatization. Similarly, the 'young reformers' were divided over the imposition of oil export tariffs in January 1992. Those in favour of the tariff included financial agencies such as the Ministry of Finance and the Ministry of Foreign Economic Relations, whereas Mintopenergo objected to the tariff.

44. The other two appointees were Viktor Shumeyko (first deputy prime minister) and Georgy Khiza (deputy prime minister). Lopukhin was not dismissed because he opposed the preferences of the 'oil generals' to create VIOCs; as mentioned above, Lopukhin was supportive of such plans. Rather, in a political tug-of-war with the Soviet-era parliament, 'some kind of sacrifice was demanded from Yel'tsin for the first painful period of price liberalisation ... and Lopukhin turned out to be the weakest link in Gaydar's government' (from author's interview with Mr Viktor Ivanenko, ex-acting chairman of the KGB and former vice-president of YUKOS, 6 February 2004, Moscow). In fact, Lopukhin's dismissal had already been demanded by the parliament in April 1992 (see Yeltin, *The Struggle for Russia*, pp. 164–167). Lopukhin's ministry not only failed to halt the decline in oil production, but his management style – he was apparently unnecessarily verbose and held unproductive meetings long into the night – also did not sit well with his peers. Lopukhin himself admitted that 'there were a lot of problems to be solved and maybe I was poorly organised' as a minister (from author's interview with Mr Vladimir Lopukhin).

45. As reported by Vasily Kononenko, 'Oil Prices in Russia to be Freed: Gas and Electricity Prices will remain Controlled', *Izvestiya*, 10 September 1992 (compiled in *Current Digest of the Post-Soviet Press*, XLIV, 36, 1992, p. 28).

46. The seminal decree 1403 was entitled 'On the Features of the Privatisation and Transformation into Joint Stock Companies of State Enterprises of Industry, Research and Production Associations of Petroleum and Petroleum Refining Industries'.

47. Quoted in 'Yel'tsin Orders Privatisation Move for Russian Oil', *Oil & Gas Journal*, 9 November 1992.

48. Author's interview with Mr Vladimir Lopukhin.
49. Author's interview with Mr Dmitriy Romanov.
50. 'Yel'tsin Orders Privatisation Move for Russian Oil'.
51. From Johnson and Kroll, 'Managerial Strategies for Spontaneous Privatisation'.
52. Ibid.
53. Chronology from Moser and Oppenheimer, 'The Oil Industry: Structural Transformation and Corporate Governance', pp. 305–306.
54. From 'Nasha kompaniya: Istoriya kompanii', the homepage of Sibneft' oil company and available at http://www.sibneft.ru.
55. This 'Concept' also called for the creation of a national oil company to represent Russia's interests in production-sharing agreements and to coordinate the oil sector as a whole.
56. From author's interview with Mr Viktor Ott.
57. See the classic statement by William A. Niskanen, *Bureaucracy and Representative Government* (Chicago: Aldine, 1971).
58. David Colander, 'Introduction', in *Neoclassical Political Economy: The Analysis of Rent Seeking and DUP Activities*, ed. David Colander (Cambridge, Massachusetts: Ballinger, 1984), p. 5.
59. Boyko, Shleifer and Vishny, *Privatising Russia*, p. 11.
60. The use of production-sharing agreements, however, was opposed by other bureaucratic interests such as the finance, customs and tax authorities who were reluctant to lose revenue that could be extracted from foreign investors. See Paul Chaisty, *Legislative Politics and Economic Power in Russia* (Basingstoke: Palgrave Macmillan, 2006), pp. 174–192.
61. Author's interview with Mr Yuriy Shafranik, ex-governor of Tyumen' oblast' and former Minister of Fuel & Energy of the Russian Federation, 13 February 2004, Moscow.
62. Shafranik recalled that he objected so strenuously to the loans-for-shares scheme that 'you wouldn't find anywhere any signature of mine on such documents' (ibid.)
63. Sivak, 'Where Will the Heavy Pressure of the Ministry of Fuel and Energy Lead?'. See also 'Russian Energy Ministry May be Dropped Soon', *Platt's Oilgram News*, 5 May 1992.
64. Tat'yana Koshkareva and Rustam Narzikulov, 'Liberaly delyat neftyanuyu sobstvennost', *Nezavisimaya gazeta*, 10 December 1998. Shafranik's successor, Sergey Generalov, also faced a similar accusation. See John Kenyon, 'Four State Oil Majors Sign Their Way toward Union', *Moscow Times*, 9 December 1998.
65. See works by Solnick, *Stealing the State*; Whitefield, *Industrial Power and the Soviet State*; and Johnson and Kroll, 'Managerial Strategies for Spontaneous Privatisation'.
66. Leonid Radzikhovskiy, 'Nomenklatura obmenyala "Kapital" na kapital', *Izvestiya*, 7 March 1995.
67. See the next chapter for details.
68. See Timothy Frye, *Brokers and Bureaucrats: Building Market Institutions in Russia* (Ann Arbor: University of Michigan Press, 2000).
69. Recounted in author's interview with Mr Vladimir Lopukhin.
70. See 'Model Russian Republic Oil Law Taking Shape', *Oil & Gas Journal*, 11 November 1991.

71. See Robert H. Bates, 'Comment', in *The Political Economy of Policy Reform*, ed. John Williamson (Washington DC: Institute for International Economics, 1994), pp. 31–33; and Bates and Krueger, 'Generalisations from the Country Studies', pp. 463–467.

72. The use of the term 'shock therapy' to describe reforms undertaken by Gaydar is not without controversy, and as such, is referred to in inverted commas. For some scholars, the term is a misnomer since 'shock therapy' was never fully implemented as a result of political pressures. See, for example, Aslund, 'Why Has Russia's Economic Transformation Been so Arduous?'; and Jeffrey D. Sachs, 'Why Russia Has Failed to Stabilise', in *Russia's Economic Transformation in the 1990s*, ed. Anders Aslund (Washington DC: Pinter, 1997).

73. As quoted by Hoffman, *The Oligarchs*, p. 178. Shevtsova makes a similar comment that for Yel'tsin, power was 'more important than life itself' (p. 270) and that the 'goal of power justified the means' (p. 271). From Lilia Shevtsova, *Yeltsin's Russia: Myths and Reality* (Washington DC: Carnegie Endowment for International Peace, 1999). Similar observations are made by Hough, *The Logic of Economic Reform in Russia*, p. 129; and Lukin, *The Political Culture of the Russian 'Democrats'*, p. 293.

74. Boris Yel'tsin, *The Struggle for Russia*, trans. Catherine A. Fitzpatrick (New York: Random House, 1994), p. 126.

75. For an insightful analysis of Yel'tsin's personality, preferences and the origins of his management style, see George W. Breslauer, *Gorbachev and Yeltsin as Leaders* (Cambridge: Cambridge University Press, 2002).

76. Yel'tsin, *The Struggle for Russia*, p. 126. In another example of Yel'tsin's predilection for haste, he insisted that 'it was paramount to reject the path of the communist economy in one fell swoop'. From Boris Yel'tsin, *Midnight Diaries*, trans. Catherine A. Fitzpatrick (New York: Public Affairs, 2000), p. 105.

77. See the next chapter for a discussion of GKI's legal prerogatives vis-à-vis other bureaucratic agencies.

78. Under Yel'tsin, the only head of GKI to question privatization was Vladimir Polevanov (November 1994–January 1995), who suggested that there was a need to renationalize some of the assets that had been privatized.

79. Dunleavy proposed this perspective of a rational, bureau-shaping bureaucrat as an alternative to Niskanen's budget-maximizing model. See Patrick Dunleavy, *Democracy, Bureaucracy and Public Choice: Economic Explanations in Political Science* (Hemel Hempstead: Harvester-Wheatsheaf, 1991); and Patrick Dunleavy, 'Explaining the Privatisation Boom: Public Choice versus Radical Approaches', *Public Administration* 64, no. 1 (1986): 13–34.

80. An example of a GKI official who was perhaps less ideologically committed than many of his peers is Al'fred Kokh, a former chairman, who frankly admitted that he valued monetary rewards over ideological satisfaction. See excerpts of interviews with Kokh in David Satter, *Darkness at Dawn: The Rise of the Russian Criminal State* (New Haven: Yale University Press, 2003), pp. 34–35. Kokh's preferences were also confirmed in the author's interview with Mr Jonathan Hay, former head of the Harvard Institute for International Development (Moscow office), 17 November 2004, London.

81. Andrey Kolesnikov, *Neizvestnyy Chubays: stranitsy iz biografii* (Moscow: Zakharov, 2003), p. 141. A similar motivation was also evident among Chubays's colleagues, one of whom noted that 'we needed to totally change

the system. Who could do this but us? . . . We had to try and make the first and, maybe, last attempt . . . This was the small window of opportunity when you could change your country'. From author's interview with Dr Andrey Konoplyanik, former deputy minister of Fuel & Energy of the Russian Federation, 28 June 2004, Brussels.

82. Quote by Chubays as cited in Brady, *Kapitalizm*, p. 64.
83. Freeland, *Sale of the Century*, p. 52.
84. The discussion here will be of a general nature. For specific examples of general managers who influenced the process and the resources at their disposal, see the subsequent case studies on YUKOS and Slavneft'.
85. These issues will only be mentioned in passing as the focus of the book is on privatization, rather than liberalization, of the oil industry. For more information on liberalization reforms, see *Energy Policies of the Russian Federation*, pp. 23–28; Pleines, 'Corruption and Crime in the Russian Oil Industry'; and Iskander Seifulmulukov and Eric Whitlock, 'Deregulation in the Russian Oil Industry', *Radio Free Europe/Radio Liberty Research Reports*, 27 August 1993, pp. 58–63.
86. See Lane, 'The Political Economy of Russian Oil', p. 104.
87. As presented in 'Znaniya: Kto my'.
88. Klebnikov, *Godfather of the Kremlin*, pp. 194–208; and 'Governor, Mayor Fight over Tax Revenue in Omsk', *Russian Regional Report*, 3, no. 44, 5 November 1998.
89. For more information on the ambition of political elites, see Bruce Kellison, 'Tiumen, Decentralisation and Centre–Periphery Tension', in *The Political Economy of Russian Oil*, ed. David Lane (Lanham: Rowman and Littlefield, 1999).
90. The proposal for the Corporation was quickly abandoned in view of fierce opposition from general directors of oil-production associations located in Tyumen'; they wanted to run their associations independently of the Corporation.
91. Shafranik and Kryukov, *Zapadno-Sibirskiy Fenomen*, p. 54.
92. In his capacity as president of the Russian Federation, Yel'tsin issued a decree in September 1991 which recognized the right of Tyumen oblast' to use its own natural resources for economic development.
93. It was predicted that oil prices would rise by five to sevenfold, far outstripping the threefold increases in prices of other goods. See 'Bigger Plunge Could Lie Ahead for Russian Oil Production'.
94. 'Russia Too Sluggish on Oil Privatisation', *Oil & Gas Journal*, 23 November 1992.
95. The first phase from 1992 to 30 July 1994 was known as mass or voucher privatization, where the main objective was to create as many shareholders and as quickly as possible. The legislation for the second phase of privatization was approved by presidential decree 1535 on 22 July 1994 and is known as the 'Basic guidelines of the state programme of privatization of state and municipal enterprises in the Russian Federation after July 1, 1994'. Its aim was to improve the efficiency of individual enterprises by mobilizing investments. See for instance, Igor' Karpenko, 'Rossiya nachinaet denezhnuyu privatizatsiyu', *Izvestiya*, 7 March 1995.
96. Appendix 2 summarizes the results of the loans-for-shares auctions.

97. Reddaway and Glinski, *The Tragedy of Russia's Reforms*, p. 480.
98. Kokh, *The Selling of the Soviet Empire*, p. 177.
99. From *Russian Economic Trends*, vol. 4, no. 3 (1995), p. 94.
100. As cited by Kokh, *The Selling of the Soviet Empire*, p. 100.
101. Author's interview with Mr Sergey Molozhaviy, former deputy minister of GKI, 10 February 2004, Moscow. The link between the ban on oil sales and loans-for-shares is also noted by Aleksandr Bekker, 'Trench Warfare on the Main Front', *Segodnya*, 29 July 1995 (compiled in the *Current Digest of the Post-Soviet Press*, XLVII, 31, p. 17), Petr Kir'yan, 'Ekonomicheskiy analiz "privatizatsiya": piat let – ne srok', *Ekspert*, 13 November 2000; and Kokh, *The Selling of the Soviet Empire*, pp. 100–105.
102. The average monthly inflation rates between 1992 and 1994 were 127% (1992), 73% (1993) and 26% (1994). See *Transition Report 1999* (London: European Bank for Reconstruction and Development, 1999), p. 261. On the use of non-cash sources of financing the federal budget, see Brian Aitken, 'Falling Tax Compliance and the Rise of the Virtual Budget in Russia', in *IMF Staff Papers* (Washington DC: International Monetary Fund, 2001).
103. Cited by Ira W. Lieberman and Rogi Veimetra, 'The Rush for State Shares in the "Klondike" of Wild East Capitalism: Loans-for-Shares Transactions in Russia', *The George Washington Journal of International Law and Economics* 29, no. 3 (1996), footnote 9, p. 743.
104. Quote by a Western investment banker as cited by Michael Comerford, 'Oil Giants' Stock Offering "Close to Free"', *Moscow Times*, 19 October 1995.
105. Data on the market capitalization of Surgutneftegaz and LUKoil are as at October 1995. These were the only two holding companies in the oil sector that were listed on the stock market, the bulk of which comprised individual oil-producing subsidiaries of holding companies. See 'Ekspert 200: Reyting krupneyshikh predpriyatiy Rossii – 1995 god', *Ekspert*, available at http://www.expert.ru/expert/ratings/exp200/exp20095/w1.htm.
106. See Robert Corzine and Nicholas Denton, 'ARCO pays', *Financial Times (London)*, 30 March 1996; Kryukov and Moe, The Changing Role of Banks in the Russian Oil Sector, p. 24; and Jane Upperton, 'ARCO's LUKoil Stake Is 1st FSU Project', *Platt's Oilgram News*, 31 August 1995.
107. See Gideon Fireman, 'British Invasion: Two Big Russian Plays', *Platt's Oilgram News*; and Joel Kibazo and Martin Brice, 'BP in demand', *Financial Times (London)*, 18 November 1997.
108. See Roland Nash and Dirk Willer, 'Share Prices in Russia: The Reasons for Undervaluation', *Russian Economic Trends*, 4, no. 2 (1995): 111–126.
109. Figures cited by Rudiger Ahrend and William Tompson, 'Fifteen Years of Economic Reform in Russia: What Has Been Achieved? What Remains to be Done?', in *OECD Economics Department Working Papers* (Paris: OECD Publishing: 2005), p. 32.
110. Quote by a Western research analyst as cited by Michael Comerford, 'Rapid Oil Privatisation May Saturate Markets', *Moscow Times*, 28 October 1995.
111. Detailed figures are cited in Carsten Sprenger, 'Ownership and Corporate Governance in Russian Industry: A Survey', in *European Bank for Reconstruction and Development Working Paper* (Paris: January 2002), Table 3, p. 6.
112. Quote by Potanin as cited by Brady, *Kapitalizm*, p. 142.

113. See Mikhail Loginov, 'Zayavlenie bankov Kluba "Nadezhnost"': Drugie banki sdelali pravitel'stvu drugoe predlozhenie', *Kommersant'-Daily*, 6 April 1995; Dmitriy Volkov, 'Major Banks of Russia Warn the Government, Suggest that the Consortium Share', *Segodnya*, 6 April 1995 (compiled in *Current Digest of the Post-Soviet Press*, XLVII, 14, 1995, p. 10).

114. For an account of the work of the Treuhandanstalt with regard to managing state-owned enterprises in east Germany, see Mark Cassell, *How Governments Privatise: The Politics of Divestment in the United States and Germany* (Washington DC: Georgetown University Press, 2002).

115. For a detailed study of the development of the banking sector in Russia, see Johnson, *A Fistful of Rubles*.

116. Estimated by Thane Gustafson, *Capitalism Russian-Style* (Cambridge: Cambridge University Press, 1999), p. 84.

117. See Shleifer and Treisman, *Without a Map*, pp. 68–70. This interpretation of the introduction of treasury bonds has been challenged by Juliet Johnson, who argues that treasury bonds were introduced as a non-inflationary method for financing the budget deficit (see Johnson, *A Fistful of Rubles*, p. 123, footnote 80). Whatever the aim of treasury bonds, the point that is undisputed is that the loans-for-shares emerged as a new money-making activity for selected banks.

118. Kryukov and Moe, *The Changing Role of Banks in the Russian Oil Sector*, p. 22.

119. Quote from a spokesman of Menatep bank, as cited by Andrey Bagrov, 'Privatizatsiya: Goskomimushchestvo vspomnilo vkus pobed', *Kommersant'-Daily*, 27 September 1995.

120. According to the GKI plan of August 1995, 15% of state-owned LUKoil stock was to be included in the loans-for-shares scheme. See Svetlana Lolayeva, 'State Property Committee Expects to Fulfill the Plan for Revenues from Privatisation', *Segodnya*, 17 August 1995 (compiled in *Current Digest of the Post-Soviet Press* XLVII, 33, 1995, p. 18).

121. William Tompson, 'Privatisation in Russia: Scope, Methods and Impact', (Birbeck College, University of London, 2004), p. 7.

122. Cited in Robert J. Brym, 'Voters Quietly Reveal Greater Communist Leanings', *Transition*, 8 September 1995.

123. From Shevtsova, *Yeltsin's Russia: Myths and Reality*, p. 139.

124. Quote by Yuriy Ivanov, as reported in 'Kak kommunisty sobirayutsya ispravlyat' grekhi demokratov', *Izvestiya*, 15 February 1996.

125. Yegor Gaidar, *Days of Victory and Defeat*, trans. Jane Ann Miller (Seattle: University of Washington Press, 1999), p. 287.

126. Quote by Gaydar from Freeland, *Sale of the Century*, p. 163.

127. During the first half of the 1990s, Russian businesses did not directly own media holdings. The tycoons preferred to pay for advertising or act as unofficial sponsors of media companies. It was only after Yel'tsin's re-election that the banks and oil companies started acquiring media-related assets. See Laura Belin, 'The Fall and Rise of State Power over the Russian Media, 1995–2001' (D.Phil. thesis, Oxford University, May 2002). The only exception was Berezovskiy: although ORT was still 51% state-owned, by autumn of 1995, he and his partners controlled 36% of the votes. Moreover, he wielded more effective control than the state representatives at ORT. See Klebnikov, *Godfather of the Kremlin*, p. 160.

128. Detractors include Matt Bivens and Chrystia Freeland, who reject the coalition-building theory as an attempt at *ex post-facto* justification. Instead, Freeland argues that loans-for-shares was simply 'a vehicle to deliver valuable state companies to the future oligarchs' (Freeland, *Sale of the Century*, p. 172) who preyed on the state's weakness; Bivens views loans-for-shares as the 'theft of natural resource companies, organised by Chubais for friends of Chubais'. (from Matt Bivens, 'Chubais and the Privatisation of History', *Moscow Times*, 26 April 2004).

129. See, for example, William B. Flemming, 'Business–State Relations in Post-Soviet Russia: The Politics of Second Phase Privatisation, 1995–1997' (M.Phil. thesis, Oxford University, 1998). With regard to this question, Primakov has argued that Yel'tsin was generally not capable of such decisions, but that his coterie of advisors dubbed 'The Family' were more than capable of strategic thinking, and that they used Yel'tsin to enhance their own wealth and influence over policy. See Yevgeny Primakov, *Russian Crossroads: Toward the New Millennium*, trans. Felix Rosenthal (New Haven: Yale University Press, 2004), especially pp. 279–313.

130. From Moser and Oppenheimer, 'The Oil Industry', footnote 11, p. 312.

131. For more details, see ibid., pp. 311–312. Upon learning of the intention by the management of Rosneft' to bid in the trust auctions, the managers of Surgutneftegaz and ONEKSIMbank (the latter was the organizer for the auctions) threatened to 'place a comma in the wrong place in the payment document'. Indeed they said, 'it is our bank, isn't it? We are an organising bank, so that is what we shall do'. (recounted in author's interview with Mr Aleksandr Putilov).

132. From an account by Lieberman and Veimetra, 'The Rush for State Shares in the Klondike of Wild East Capitalism', p. 748.

133. See Sergey Leskov, 'Dogovornye matchi byvayut ne tol'ko o sporte: ocherednoy auktsion neftyanykh kompanii', *Izvestiya*, 29 December 1995.

134. See Kryukov and Moe, *The Changing Role of Banks in the Russian Oil Sector*, p. 38.

135. Ibid., p. 33.

136. Details of the accusation are recounted by Jane Upperton, 'Russia's Latest Share Brings Fresh Criticism', *Platt's Oilgram News*, 2 July 1997.

137. Tatneft' is not included among the top five VIOCs mentioned here as it is partly owned by the regional authorities.

138. Anatoliy Chubays, 'V poiskakh sobstvennogo puti', in *Privatizatsiya po-rossiyskiy*, ed. Anatoliy Chubays (Moskva: Vagrius, 1999), p. 323.

139. Quote by Sergey Markov, director of the Centre for Political Studies, in Gregory Feifer, 'The Kremlin's Big Sell-Off', *Moscow Times*, 19 December 2000.

140. Quote by Vladimir Malin of the Federal Property Fund from 'Russian Government Sells Onako State Oil Company for $1.08 Billion', *The Associated Press*, 19 September 2000.

141. For example, Berezovskiy was not invited to attend a meeting between the newly elected president and leading businessmen held in June 2000, while his business acquisitions were subjected to probes by the Prosecutor General's office around this time. For more on the politics behind ONAKO's sale, see Feifer, 'The Kremlin's Big Sell-Off'.

142. Cited by Eduard Gismatullin, 'One Horse Race for TNK Stake', *Moscow Times*, 6 November 1999.
143. One of Berezovskiy's most famous mantras relates to the fact that one does not have to buy an enterprise in order to control it. One can resort to 'privatisation of profits', 'privatisation of property' or 'privatisation of debts', as quoted by Klebnikov, *Godfather of the Kremlin*, p. 170.
144. The Russian government owned 75% of Slavneft', while the remainder was held by the Belarussian government. An in-depth analysis of the privatization of Slavneft' is at Chapter 4.
145. For a discussion on the role of change teams and management of reforms, see Nelson, 'Conclusion', p. 335; and Williamson, 'In Search of a Manual for Technopols', pp. 25–26.
146. For more on the strategic nature of Yel'tsin's ever-changing governments, see Reddaway and Glinski, *The Tragedy of Russia's Reforms*, pp. 243–244 (the 'seesaw' system); and Breslauer, *Gorbachev and Yeltsin as Leaders*, pp. 251–254.
147. Yel'tsin, *Midnight Diaries*, p. 104.
148. Lilia Shevtsova, 'From Yel'tsin to Putin: The Evolution of Presidential Power', in *Gorbachev, Yeltsin and Putin: Political Leadership in Russia's Transition*, ed. Archie Brown and Lilia Shevtsova (Washington DC: Carnegie Endowment for International Peace, 2001), pp. 69–70. A similar perspective is by Medvedev, *Post-Soviet Russia*, p. 238.
149. Putin, 'Russia at the Turn of the Millennium'.
150. Andrew Jack, *Inside Putin's Russia* (London: Granta, 2004), p. 334.
151. Nataliya Gevorkyan, Natalya Timakova and Andrei Kolesnikov, *First Person: An Astonishingly Frank Portrait by Russia's President Vladimir Putin*, trans. Catherine A. Fitzpatrick (New York: Public Affairs, 2000), p. 193.
152. Quote by Sergey Markov, director of Centre for Political Studies in Moscow, as cited by Feifer, 'The Kremlin's Big Sell-Off'.
153. See Putin's advice to leading businessmen at July 2000 meeting as reported by Igor' Trefilov, 'Skromnee, tishe, dal'she: Kakim dolzhen byt' oligarkh novogo vremeni', *Segodnya*, 29 July 2000. The observation that Putin appears to favour certain businessmen over others is also noted by Shevtsova, 'From Yel'tsin to Putin', footnote 48, p. 111.
154. Brady, *Kapitalizm*, p. 139.
155. Quote by Aleksandr Kazakov, Chairman of GKI, in 'Interv'yu Aleksandra Kazakova: Vse dolzhno idti estestvennym putem', *Kommersant'-Daily*, 13 February 1996.
156. From Yitzhak M. Brudny, 'Continuity or Change in Russian Electoral Patterns? The December 1999–March 2001 Election Cycle', in *Contemporary Russian Politics: A Reader*, ed. Archie Brown (Oxford: Oxford University Press, 2001). McFaul has come up with a different set of numbers. According to his analysis, pro-reform groups received 43% of votes in 1993, but only 38.2% in 1995. In contrast, their opponents received 42.8% of the votes in 1993 and 52.8% in 1995. See Michael McFaul, *Russia between Elections: What do the 1995 Results Really Mean?* (Washington DC: Carnegie Endowment for International Peace, 1996), p. 3. The discrepancies with Brudny's figures are due to the way in which parties are classified. In any case, the point is the same: that the Communist Party and its allies had increased their share of the vote at the expense of liberal parties.

157. The figures are from Shevtsova, *Yeltsin's Russia: Myths and Reality*, p. 156.
158. See, for instance, 'The Communist Comeback', *New York Times*, 19 December 1995; 'Communists and Nationalists Do Better than Expected', *New York Times*, 19 December 1995; Jerry Hough, Evelyn Davidheiser and Susan Goodrich Lehmann, *The 1996 Presidential Election* (Washington DC: Brookings Institution Press, 1996); and Peter Reddaway, 'Red Alert', *The New Republic*, 29 January 1996. For a dissenting opinion regarding the balance of forces in the country, see Michael McFaul, *Russia's Unfinished Revolution: Political Change from Gorbachev to Putin* (Ithaca: Cornell University Press, 2001), pp. 286–287.
159. Aleksandr Bekker, 'Gibkaya politika GKI', *Segodnya*, 12 April 1996.
160. See, for example, 'Khodorkovskiy stal kumirom delovoy molodezhi', *Nezavisimaya gazeta*, 21 July 2003; Jason Bush, 'Russia: Why Business is Rushing into Politics', *Business Week*, 8 December 2003; Anna Politkovskaya, *Putin's Russia*, trans. Arch Tait (London: Harvill Press, 2004), p. 276; Natal'ya Arkhangel'skaya, 'Komitet po delam oligarkhov', *Ekspert*, 19 May 2003; and Anatoliy Kostyukov, 'Ten' YUKOSa nakryla i npavykh, i levykh', *Nezavisimaya gazeta*, 9 December 2003.
161. Respectively, the quotes are cited in Arkhangel'skaya, 'Komitet po delam oligarkhov'; and Natal'ya Zakharchk, 'Serzhantov khvatit na vsekh', *Moskovskiye novosti*, 22 July 2003.
162. From Susan B. Glasser and Peter Baker, 'Two Visions for Russia and One Battle of Wills: Oilman has History of Defying Putin', *The Washington Post*, 5 November 2003; and Francesca Mereu, 'Yukos Takes a Bite out of Yabloko's Party List', *Moscow Times*, 3 December 2003.
163. The party list of the Communists included businessmen linked to TNK while that of the Union of Rightist Forces included executives linked to the electricity monopoly, Unified Energy Systems and its president Chubays.
164. Estimate cited above in Mereu, 'Yukos Takes a Bite out of Yabloko's Party List'.
165. This prospect of an 'oligarchic coup' was contained in a notorious report 'V Rossii gotovitsya oligarkhicheskiy perevorot', (Moscow: Sovet po Natsional'noy Strategii, 26 May 2003, available at http://www.utro.ru/articles/2003/05/26/201631.shtml.) As ringleader, Khodorkovskiy was mentioned 10 times and YUKOS 25 times, Abramovich and Sibneft' were given 9 and 22 mentions and Fridman and Al'fa received 4 and 10 mentions. See Vladimir Pribylovsky, 'What's the Scandal All About?', *Moscow Times*, 11 June 2003.
166. For a discussion of the primary cause of the YUKOS affair, see the following chapter.
167. Mark Zavadskiy, "Vykovyrivat' i mochit", *Yezhenedel'nyy zhurnal*, 3 November 2003.
168. See Vladislav Borodulin, 'Privatizatsiya v Rossii: Boris El'tsin vernul VPK v lono byudzheta', *Kommersant'-Daily*, 26 December 1995.
169. The term is from Stephen Holmes, 'Superpresidentialism and its Problems', *East European Constitutional Review* 3, no. 1 (1994), p. 124.
170. See the discussion in Chapter 4.
171. Quote by Viktor Semago as cited by Andrei Jr Zolotov, 'The Art of the Deal', *Moscow Times*, 19 August 2000. See also Yekaterina Zapodinskaya, 'V Gosdume brali vzyatki po preyskurantu', *Kommersant'-Daily*, 16 March 2000.

172. As cited in Christopher Thomas Speckhard, 'The Ties that Bind: Big Business and Centre–Periphery Relations in the Russian Federation' (D.Phil. thesis, University of Texas at Austin, December 2004). The key actors included the long-serving governor of Samara, Konstantin Titov and YUKOS's vice-president Viktor Kazakov.

173. Author's interview with Mr Sergey Molozhaviy.

174. Paul Chaisty and Petra Schleiter, 'Productive but Not Valued: The Russian State Duma, 1994–2001', *Europe–Asia Studies*, 55, no. 5 (2002), p. 714.

175. See Maxim Ryzhov, 'Energichnaya Gosduma', *Neft' i Kapital*, 2000; and Raj M. Desai and Itzhak Goldberg, 'The Politics of Russian Enterprise Reform: Insiders, Local Governments, and the Obstacles to Restructuring', *The World Bank Research Observer* 16, no. 2 (Fall 2001): 219–240.

176. As cited by Yuriy Zarakhovich, 'Meet the Second Richest Man in Russia', *Time Europe*, 2 December 2002.

177. For more examples of oil-executives-turned senators, see Aleksey Makarkin, 'Senatory ot nefti', *Neft' i Kapital*, 2002. The influence of oil and other business groups on regional elections is the subject of an article in 'Regional Leaders and their Corporate Sponsors', *Radio Free Europe/Radio Liberty: Russian Political Weekly*, 23 January 2003 (available at www.rferl.org/rpw). However, the oil lobby is not a united group and the interests of individual companies often clash in the regions.

3 Case Study of YUKOS

1. For example, the name and role of Sergey Muravlenko is largely absent in accounts by Lane and Seifulmulukov, 'Structure and Ownership'; Freeland, *Sale of the Century*; Hoffman, *The Oligarchs*; Goldman, *The Piratization of Russia*; and Fortescue, *Russia's Oil Barons and Metal Magnates*.

2. See 'Hostile Takeover, Russian Style', *Forbes*, 21 November 1994; and 'Zaderzhany podozrebaemye v ubiystve krupnogo biznesmena', *Kommersant'-Daily*, 26 October 1995.

3. All kinds of spurious reasons were given for such requests. For instance, Russia's vice-president Aleksandr Rutskoi requested a certain quota of oil to be given to a specific company to commemorate the 500th anniversary of the founding of America by Christopher Columbus (from author's interview with Mr Lev Churilov). Such requests continued to be made during Gaydar's tenure as acting prime minister, as recounted by Gaidar, *Days of Victory and Defeat*, pp. 122–123.

4. Gaidar, *State & Evolution*, p. 59. A similar argument is made by Solnick, *Stealing the State*.

5. Author's interview with Dr Rair Simonyan (in Oxford).

6. Author's interview with Mr Andrey Shtorkh, former vice-president of Slavneft', 1 October 2003.

7. 'Press Conference with the President of the LUKoil Company', *Official Kremlin International News Broadcast*, 5 April 1995.

8. Viktor Muravlenko was in charge of the Soviet Union's largest oil enterprise, Glavtyumenneftegaz, from 1965 till his death in 1977. As a mark of respect, a drill ship and town in Tyumen' were named in his honour. His contributions

to the Soviet oil industry are recounted by Alexander Atanatsky, 'From the Glorious Ranks of the Pioneers', *Oil of Russia*, no. 4, 2005; and by Tchurilov, *Lifeblood of Empire*.

9. For further details of oil production by enterprise, see *Energy Policies of the Russian Federation*, p. 97. Figures on oil reserves may be found in the same publication (p. 106).

10. Author's interview with Mr Viktor Ivanenko, ex-acting chairman of the KGB and former vice-president of YUKOS, 6 February 2004, Moscow; and author's interview with Mr Vladimir Lopukhin.

11. Foreign consultants (Daiwa Europe and Bankers Trust) hired by the state had in fact proposed that Yuganskneftegaz be merged with Yaroslav refinery, but the suggestion was never considered by YUKOS's management since the refinery's capacity was much too small. Details of the proposal may be found in 'Blueprint for Reform: New Policies and Structures for the Russian Oil Industry', (New York: Petroleum Intelligence Weekly, 1992), pp. 56, 58.

12. Moser and Oppenheimer, 'The Oil Industry', p. 306.

13. Author's interview with Mr Aleksandr Putilov.

14. Author's interview with Mr Viktor Ott.

15. The figure is from Ol'ga Kryshtanovskaya, 'Finansovaya oligarkhiya v Rossii', *Izvestiya*, 10 January 1996. On the issue of the extent of elite change, the elite adaptation thesis claims that the Soviet *nomenklatura* have simply transformed themselves into the new political and economic elites. See, for instance, Olga Kryshtanovskaya and Stephen White, 'From Soviet Nomenklatura to Russian Elite', *Europe–Asia Studies* 48, no. 5 (1996): 711–733; and Hellmut Wollman, 'Change and Continuity of Political and Administrative Elites in Post-Communist Russia', *Governance*, no. 6 (1993): 326–340. In contrast, supporters of the elite competition thesis argue that intra-elite conflicts along generational, ideological or technocratic lines provide a more complex explanation of developments in Russia. See Iosif Diskin, *Rossiya: Transformatsiya i elity* (Moscow: Eltra, 1995); David Lane, 'The Transformation of Russia: The Role of the Political Elite', *Europe–Asia Studies* 48, no. 4 (1996): 535–549; and Ovsei Shkaratan and Yuriy Figatner, 'Starye i novye khozyaeva Rossii', *Mir Rossii* 1, no. 1 (1992): 67–90.

16. The estimate by the World Bank is from Charles P. McPherson, 'Policy Reform in Russia's Oil Sector', *Finance & Development: A Quarterly Magazine of the IMF*, June 1996. The figure on the stock of foreign direct investment between 1992 and 1995 is from *World Investment Report 2003*, (New York: United Nations, 2003), p. 260. In comparison, during the same period, smaller countries such as the Czech Republic and Poland attracted FDI of US$7.3 billion and US$7.8 billion, respectively.

17. Two separate loans to various production enterprises in the oil sector were extended by the World Bank in 1993 and 1995. Yuganskneftegaz received US$190 million in 1995, but later cancelled part of the loan due to the conditions attached. For a review of the Bank's policy in Russia, see Bobylev and Cukrowski, 'Russia: Bank Assistance for the Energy Sector'.

18. Author's interview with Mr Viktor Ivanenko.

19. Cited in Andrey Fedorov, 'Kadrovye peremeny v YUKOS: Smena rukovodstva na Novokuybyshevskom NPZ', *Kommersant'*, 18 April 1995; 'Arbitrazhnye sudy', *Kommersant'*, 17 November 1993; and author's interview with Mr Viktor

Ivanenko. Tarkhov, the Soviet-era chairman of Samara oblast', was removed by Yel'tsin in 1991 for siding with the leaders of the August 1991 coup. Tarkhov then quickly exchanged political power for economic rights when he became the general manager of Samara's NovoKuybyshev oil refinery.

20. See Forbes, 'Hostile Takeover, Russian style'. One of Khodorkovskiy's first task after purchasing YUKOS was to dispatch 300 of his best security personnel to Siberia to physically take over the company's wells and refineries, many of which were in the hands of criminals. He also visited every single financial controller and accountant in the subsidiary companies to let them know that he was their boss. See the account by Hoffman, *The Oligarchs*, pp. 445–446.

21. Author's interview with Dr Rair Simonyan (in Moscow).

22. For a review of the relationship between company law on governance, see Katharina Pistor, 'Company Law and Corporate Governance in Russia', in *The Rule of Law and Economic Reform in Russia*, ed. Katharina Pistor and Jeffrey D. Sachs (Boulder: Westview Press, 1997).

23. Andrey Fedorov and Aleksandr Tutushkin, "Sozdaetsya AO 'YUKOS-Samara': YUKOS ustal zhdat'", *Kommersant'-Daily*, 2 June 1995. Muravlenko's non-confrontational style of management was confirmed in interviews conducted with his former colleagues (he stepped down as Chairman of YUKOS in 2003).

24. Author's interview with Mr Viktor Ivanenko.

25. This is an argument for concentrated ownership, as cited in J. David Brown and John S. Earle, 'Evaluating Enterprise Privatisation in Russia', *Russian Economic Trends* 8, no. 3 (1999), p. 26. Indeed, there appears to be a positive correlation between an improvement in the firm's performance and the consolidation of majority stakes by manager-owners. See, for instance, Boone and Rodionov, 'Rent Seeking in Russia and the CIS'; and Sergei Guriev and Andrei Rachinsky, 'Ownership Concentration in Russian Industry', (Background paper for Russia CEM 2003: The World Bank and Princeton University, March 2004).

26. Khodorkovskiy also doubted that the state would ever divest its shareholdings in the oil companies. In his own words, 'I could never believe that the state would sell oil' (Hoffman, *The Oligarchs*, p. 302.) Indeed, only in 1995 did he begin to believe that it was possible to acquire oil companies.

27. See Aleksey Sukhodoev and Aleksandr Tutushkin, 'Otstavka glavy 'Yuganskneftegaza': Neftyanogo generala ubrali po sostoyaniyu finansov', *Kommersant'-Daily*, 26 October 1995.

28. Author's interview with Mr Viktor Ivanenko. YUKOS, and in particular, Yuganskneftegaz, was one of the largest tax debtors in the country at that time, along with AvtoVaz, Nizhnevartovskneftegaz and SIDANKO, as noted in 'Enterprises and Banking', *Russian Economic Trends*, no. 3 (1998), Table 3, p. 36.

29. This transfer of Purneftegaz effectively dismembered SIDANKO as a VIOC since the company was left with a severe shortage of crude petroleum and excess refining capacity.

30. Figures on shareholdings are provided by Boone and Rodionov, 'Rent Seeking in Russia and the CIS', p. 9.

31. Author's interview with Mr Sergey Generalov, former vice-president of YUKOS and ex-minister of Fuel & Energy of the Russian Federation, 30 September 2003, Moscow.

32. A leading newspaper even referred to him as a 'hereditary oilman'. See Dmitriy Butrin, 'Zhizn' posle YUKOSa', *Kommersant' Den'gi*, 26 July 2004.
33. Author's interview with Mr Viktor Ivanenko.
34. See Olson, *Power and Prosperity*. The adoption of a mixed strategy is also discussed by Bernard Black, Reiner Kraakman and Anna Tarassova, 'Russian Privatisation and Corporate Governance: What Went Wrong?', *Stanford Law Review* 52 (2000): 1731–1808.
35. See the argument by Raj M. Desai and Itzhak Goldberg, 'The Vicious Circles of Control: Regional Governments and Insiders in Privatised Russian Enterprises', in *World Bank Working Paper* (Washington DC: World Bank, 2000).
36. The cash privatization programme was promulgated through presidential decree 1535 of 22 July 1994 and entitled 'Basic Guidelines of the State Programme of Privatisation of the State and Municipal Enterprises in the Russian Federation after July 1, 1994'.
37. Promradtekh bank's relatively poor position in a survey of banks between 1992 and 1997 is given by Kryukov and Moe, *The Changing Role of Banks in the Russian Oil Sector*, Table 1, p. 9. The architect of the bank's initial plan was in fact Sergey Generalov, who later joined YUKOS as its vice-president.
38. Recounted in author's interview with Mr Viktor Ivanenko.
39. See the survey of banks in Kryukov and Moe, *The Changing Role of Banks in the Russian Oil Sector*, p. 9.
40. Author's interview with Mr Viktor Ivanenko.
41. They included Menatep itself, Tokobank (20% of its shares were owned by Yuganskneftegaz, a subsidiary of YUKOS), Promradtekh bank (which was used by YUKOS for its public share issue in 1994), Stolichniy Bank (which was one of the banks that provided the guarantee for front companies bidding in the investment tender for YUKOS), and major suppliers or customers of YUKOS, such as Nafta-Moskva, Roskontrakt, Splav and Kurganmashzavod. The Russian origin of these companies was also emphasized by Khodorkovskiy in an interview with Yuriy Katsmanu, 'Interv'yu rukovoditelya banka 'Menatep'', *Kommersant'-Daily*, 1 December 1995.
42. Author's interview with Mr Sergey Generalov.
43. For a report on the shareholders' meeting, see Aleksandr Tutushkin, 'Sobranie aktsionerov kompanii YUKOS: Aktsionery zafiksirovali svoy patriotizm v ustave', *Kommersant'-Daily*, 26 December 1995.
44. Quote by Viktor Ivanenko, vice-president of YUKOS, as cited by Yuriy Ryazhskiy, 'Spiders in the Banks: Oil, Dollars, Scandal', *Moskovskiy Komsomolets*, 30 November 1995 (compiled in *Current Digest of the Post-Soviet Press* XLVII, no. 49, 1995, pp. 8–9). Al'fa bank supposedly intended to turn over part of YUKOS's shares to an American company, the Davis Petroleum Company.
45. Attributing human agency to a collective organization, such as a firm, is a common practice in the field of microeconomics, as discussed in a previous chapter. Prior to the YUKOS affair, Khodorkovskiy directly owned 9.5% of Menatep and controlled another 50% through a special trust arrangement. Menatep, in turn, owned 61% of YUKOS. At the end of 2004, Khodorkovskiy resigned from his positions in YUKOS and Menatep and transferred his 59.5% stake in Menatep to Nevzlin, who lives in Israel. See Irina Reznik,

Yuliya Bushueva, Aleksandr Bekker and Tat'yana Egorova, 'Khodorkovskiy vyshel', *Vedomosti*, 12 January 2005.

46. See 'Russian State Buys into Commercial Banks', *BBC Summary of World Broadcasts*, 19 May 1995.

47. This is not to say that large firms do not engage in bribery, merely that they pay a smaller proportion out of their enormous revenues as bribes, compared to smaller firms. Examples of empirical studies to this effect include Hellman, Jones and Kaufman, 'Seize the State, Seize the Day'; Joel S. Hellman and Daniel Kaufman, 'The Inequality of Influence', (Washington DC: World Bank, 2003); and Irina Slinko, Evgeny Yakovlev and Ekaterina Zhuravskaya, 'Institutional Subversion: Evidence from Russian Regions', (Moscow: Centre for Economic and Financial Research, 2003).

48. In many of the loans-for-shares auctions, the auction organizer was also the leading candidate to acquire the collateralized shares. This was the case, for example, with Norilsk Nickel, YUKOS and SIDANKO. In other cases, the auction organizer was an ally of the leading candidate.

49. These figures are for equity capital as at 1 July 1997 as compiled by Kryukov and Moe, *The Changing Role of Banks in the Russian Oil Sector*, p. 16.

50. These allegations are repeated in author's interview with Dr Rair Simonyan (Moscow); author's interview with Mr Aleksandr Putilov; Michael Bernstam and Alvin Rabushka, *Fixing Russia's Banks: A Proposal for Growth* (Stanford: Hoover Institution Press, 1998); and Hoffman, *The Oligarchs*, pp. 316–318.

51. Foreigners were, in fact, banned from participating in auctions for 8 of the 12 companies on offer.

52. Konstantin Kagalovsky, a former vice-president of Menatep and YUKOS, admitted to drafting the condition with this intention in mind. See Freeland, *Sale of the Century*, pp. 175–176.

53. See, for example, Gleb Baranov, 'Press-konferentsiya YUKOSa i 'Menatepa': Bank nameren borot'sya za kontrol' nad kompaniey', *Kommersant'-Daily*, 10 November 1995.

54. This account is taken from Hoffman, *The Oligarchs*, p. 316.

55. Khodorkovskiy later claimed to be 'simply in shock' that Vinogradov had gone back on his word. See the previously cited interview in *Kommersant'-Daily* by Katsmanu, 'Interv'yu rukovoditelya banka "Menatep"'.

56. From 'Sovmestnoe zayavlenie kommercheskikh bankov', *Kommersant'-Daily*, 28 November 1995.

57. Ibid.

58. See 'Vedomosti: Minfin zayavlyaet', *Kommersant'-Daily*, 7 December 1995.

59. As cited in 'Zampred TSB preduprezhdaet bol'shuyu troiku', *Kommersant'-Daily*, 1 December 1995.

60. See Yaroslav Skvortsov, 'Sobranie Moskovskogo Bankovskogo Soyuza: Bankiry ponyali, shto nemnogo pereborshchili', *Kommersant'-Daily*, 5 December 1995; and 'GKI otvetil bankam: Komitet postaralsya proyasnit' vse neyasnosti', *Kommersant'-Daily*, 2 December 1995.

61. Quote by Stephen O'Sullivan, an investment banker, as cited by Poul Funder Larsen, 'Menatep Maneuvers to Give Itself YUKOS Control', *Moscow Times*, 21 November 1996.

62. They included Kagalovskiy, who was Russia's former representative to the IMF and who drafted the condition banning foreigners from YUKOS's investment

auction, Leonid Nevzlin, one of Khodorkovskiy's earliest business partners, and Aleksandr Samusev, who was a former Deputy Minister of Fuel and Energy and former Finance Minister. All three were given positions as vice-presidents in YUKOS.

63. YUKOS's efforts to impose centralized control over its subsidiaries were widely documented in the media. See, for instance, Konstantin Lange and Elena Tret'yakova, 'Soglashenie "Menatepa", YUKOSa i Samary: Neftyanaya kompaniya i bank ne obidyat oblast", *Kommersant'-Daily*, 23 March 1996; Rustam Narzikulov, 'Muravlenko budet rukovodit' "Rospromom"', *Nezavisimaya gazeta*, 24 May 1996; Aleksandr Tutushkin, 'Peremeny v kompanii YUKOSa: Syd'bonosnye resheniya YUKOS primet za granitsey', *Kommersant'-Daily*, 18 April 1996; and Aleksandr Tutushkin, 'YUKOS zavershil restrukturizatsiyu: Kompaniya posadila "dochek" na byudzhet', *Kommersant'-Daily*, 4 October 1996.

64. The figures are from Aleksandr Tutushkin, 'Sobranie aktsionerov YUKOSa: Nefyanoy kompanii stanet na tret' bol'she', *Kommersant'-Daily*, 17 September 1996. The widespread use of promissory notes and barter resulted in general demonetization and the existence of a 'virtual economy' in Russia, as explained by Clifford Gady and Barry Ickes, 'Russia's Virtual Economy', *Foreign Affairs* 77, no. 5 (1998): 53–67.

65. The decree was entitled 'On Measures to Eliminate Arrears of Joint-Stock Companies on Wages and Taxes'.

66. Alexander Radygin, 'Ownership and Control of the Russian Industry' (paper presented at the Conference on Corporate Governance in Russia, Moscow, 31 May–2 June 1999), p. 14.

67. The US$509.7 million is derived as follows: US$350 million for the investment tender in 1995, plus US$159 million for the collateral auction in 1995, plus US$700,000 for the auction of pledged shares in 1996. The valuation of the company according to market capitalization is from data compiled in an annual rating of the 200 leading companies in Russia by Ekspert magazine. The data are available at http://www.expert.ru/expert/ratings/exp200.

68. From Juliet Johnson, 'Russia's Emerging Financial–Industrial Groups', *Post-Soviet Affairs* 13, no. 4 (1997), footnote 33, p. 351.

69. Quote by Vinogradov from Poul Funder Larsen, 'Bank Battles Turning Ugly as Rivals Sling Mud', *Moscow Times*, 12 July 1996. For a more in-depth review, see Mikhail Loginov, 'Skandal vokrug izvestnogo banka: Proverka Inkombanka proizvela bol'shoy shum', *Kommersant'-Daily*, 12 July 1996.

70. See Mikhail Loginov, 'Novyy etap bankovskogo krizisa: Vot i novyy avgust podoshel', *Kommersant'-Daily*, 16 July 1996.

71. Kokh, *The Selling of the Soviet Empire*, p. 157. Kazakov's concessions on privatization were enumerated in the previous chapter.

72. See comments by analysts in Poul Funder Larsen, 'Banks Likely to Keep Shares', *Moscow Times*, 31 August 1996. The options to redeem the collateralized shares included exchanging them for government bonds and raising money to pay for their redemption by selling part of the government's hard-currency reserves or using funds previously earmarked for the reconstruction of Chechnya.

73. The 'two keys' aspect of GKI's revision to the loans-for-shares scheme was discussed in the previous chapter.

74. So concerned was Minister Shafranik about his future employment prospects that the leading oil generals agreed that an oil company, SIDANKO, would be created for him out of Rosneft, so that he would 'feel more secure and make independent decisions'. From author's interview with Mr Aleksandr Putilov.

75. For more details, see Aleksandr Bekker, 'Mintopenergo sokhranil vliyanie v TEKe: Nich'ya na promezhutochnom finishe', *Segodnya*, 15 July 1996; and Rustam Narzikulov, 'Neftyanye aktsii pod dvoynym pressom', *Nezavisimaya gazeta*, 10 April 1996.

76. Viktor Ivanov, 'Privatizatsiya v TEKe: Neft' ostaetsya v gossobstvennosti', *Kommersant'-Daily*, 12 September 1996.

77. Andrey Bagrov, 'Novaya kontseptsiya zalogovykh auktsionov: Sovet bezopasnosti polozhil glaz na pakety i banki', *Kommersant'-Daily*, 21 September 1996.

78. For a comprehensive analysis of how Khodorkovskiy consolidated YUKOS's control over its subsidiaries, see Yuko Iji, 'Corporate Control and Governance Practices in Russia', in *Centre for the Study of Economic & Social Change in Europe Working Paper* (London: June 2003). Other oil companies also engaged in lengthy battles with minority shareholders, as recounted by Black, Kraakman and Tarassova, 'Russian Privatisation and Corporate Governance'.

79. Further details about the report were noted in the previous chapter.

80. Investigations revealed later that the shell company, Baykal Finans, used to bid for Yuganskneftegaz during the auction, was owned and financed by Rosneft'. It is as yet unclear how Rosneft' paid for its purchase, with observers claiming that money was either taken from the budget or Stabilization Fund or perhaps borrowed from Russian or even Chinese banks. See Julia Latynina, 'The Yugansk Coverup Operation', *Moscow Times*, 30 December 2004; Irina Granik and Nikolay Vardul', 'Dissident: god velikogo obloma', *Kommersant'-Daily*, 29 December 2004; 'Rosneft' zalozhila sebya po-krupnomu', *Kommersant'*, 26 August 2005; and 'Kto oplatil "Yugansk"', *Vedomosti*, 3 June 2005.

81. Quote by Igor' Shuvalov, Putin's senior economic advisor, as noted in 'Putin's Economic Aide Warns Other Oil Companies May Face Back Tax Bill following Yukos', *Associated Press*, 28 October 2004 (printed in *Johnson's Russia's List* no. 8428).

82. Quote by Putin as cited by Seth Mydans, 'Putin Calls Arrests Part of a Crackdown', *International Herald Tribune*, 7 November 2003.

83. The quote by Shuvalov is from 'Putin's Economic Aide'.

84. These tax privileges enjoyed by internal-offshore zones have since been rescinded. For an account of the tax privileges enjoyed by companies operating in these zones, see Vadim Visloguzov, 'Novyy poryadok: takoy ofshor im ne nuzhen', *Kommersant'-Daily*, 18 November 2003.

85. The role of Dubov is analysed in greater detail by Aleksey Nedogonov, 'YUKOS vs SRP: Chem prodiktovana bor'ba Mikhaila Khodorkovskogo s rezhimom razdela produktsii', *RusEnergy.com*, 12 February 2003.

86. Quote by Gaydar as cited in 'A Russian Reformer Sees Pitfalls Ahead', *Business Week*, 12 December 2003.

87. For an account of the offshore companies operated by YUKOS in foreign tax havens and the actual workings of 'transfer pricing' in YUKOS, see Jeanne

Whalen, 'A Jilted Banker's View of Khodorkovsky's Empire', *Wall Street Journal*, 2 January 2003; and Iji, 'Corporate Control and Governance Practices in Russia'. For a perspective that transfer pricing was not, in fact, aimed at minimizing taxes but an accounting anomaly, see Nat Moser, 'Transfer Pricing and Calculating Russia's GDP', *Moscow Times*, 10 June 2004.

88. Estimates by independent and government analysts are from Elizabeth LeBras and Natalya Neimysheva, 'Oil Evades $9bln in Taxes', *Moscow Times*, 29 November 2000.

89. The comparison of effective tax rates for YUKOS and Surgutneftegaz in 2000 is from 'A Taxing Time for Oil', (Moscow: Renaissance Capital, 21 January 2004).

90. From 'A Taxing Time for Oil'; and Anna Raff, 'Companies Can't Wait to Pay the Tab in Russia', *Wall Street Journal*, 21 September 2004.

91. Recounted in Mariya Ignatova, 'Nalogoviki vernutsya v "Sibneft"' i "Rosneft"', *Izvestiya*, 13 May 2003.

92. Aleksey Polukhin, 'Nalogovyy prays', *Novaya gazeta*, 11 December 2003. See also Boris Grozovskiy, Svetlana Ivanova, Igor' Fedyukin and Aleksey Nikol'skiy, 'Vse biznesmeny delali eto', *Vedomosti*, 29 October 2003.

93. 'Putin Criticises Big Business', *AFP*, 9 December 2003 (printed in *Johnson's Russia's List* no. 7461). In fact, it has been noted that at that time, Russia had 'no legislation requiring tax avoidance schemes to be submitted to the tax authorities in advance for a check on their acceptability. US laws do require this, but Russia's laws, like those of most western countries, do not'. See Philip Hanson, 'Observations on the Cost of the Yukos Affair to Russia', *Post-Soviet Affairs* 46, no. 7 (2005), p. 483.

94. Claim by Fedor Chayka, 'Gosudarstvo raspravilos' s 'YUKOSom', *Izvestiya*, 30 December 2004. For an account of various solutions to settle the back tax claims, see 'YUKOS Ready to be Nationalised?', *RIA RosBusinessConsulting*, 21 June 2004 (printed in *Johnson's Russia's List* no. 8263); Catherine Belton, 'Ally Offers to Bail Out Yukos', *Moscow Times*, 26 July 2004; and Tat'yana Gurova, Maksim Rubchenko and Andrey Tsunskiy, 'Delo YUKOSa: Cherez dvoynuyu sploshnuyu', *Ekspert*, 12 July 2004.

95. Stanislav Menshikov, 'Plyaski s volkami vokrug nefti', *Slovo*, 23 July 2004. The primacy of 'property redistribution to the *siloviki*' as a motivation for the YUKOS affair was also argued by Vlad Sobell, 'Russia Post-Yukos', Daiwa Institute of Research Europe Limited, 28 July 2004 (printed in *Johnson's Russia's List* no. 8307); Shevtsova; Daniel Kimmage, 'Why the Conventional Wisdom about Russia is Wrong', *In the National Interest*, 10 January 2004 (available from http://www.inthenationalinterest.com); Yuliya Latynina, 'Navstrechu vyboram: Esli v Rossii peredel sobstvennosti – eto k vyboram', *Novaya gazeta*, 15 May 2003; and Vadim Volkov, 'The Yukos Affair: Terminating the Implicit Contract', in *PONARS Policy Memo* (Washington DC: Centre for Strategic and International Studies, November 2003).

96. The rise of the *siloviki* has been a source of concern for some scholars who warn of the danger of a militocracy and the Sovietization of the state. See Olga Kryshtanovskaya and Stephen White, 'Putin's Militocracy', *Post-Soviet Affairs* 19, no. 4 (2003).

97. A profile of leading members of the *siloviki* is given by Il'ya Bulavinov, 'Politicheskiy rasklad: koloda Rossiyskoy Federatsii', *Kommersant'-Vlast'*,

1 December 2003; and Dmitriy Kamyshev, 'Apparat: Kremlevskaya desyatka', *Kommersant'-Vlast'*, 19 April 2004. Ustinov's son Dmitriy is, in fact, married to the daughter of Igor' Sechin, widely regarded as the chief patron of the *siloviki*.

98. The dispute was over YUKOS's purchase of 19% of Yeniseyneftegaz, which was partly owned by Rosneft'. For more on Bogdanchikov's role in the YUKOS affair, see Aleksandr Losev, 'Priglashenie "YUKOSa" na kazan': v roli topora – Bogdanchikov', *Novaya gazeta*, 17 July 2003. Other points of contention between Rosneft' and YUKOS included the need for production-sharing agreements and the acquisition of Severnaya Neft' oil company by Rosneft'.

99. See Losev, 'Priglashenie "YUKOSa" na kazan"; and Elena Dikun, '"Delo Yukosa"v sude', *Moskovskiye novosti*, 21 May 2004.

100. By 1999, Putin had already accepted that Russia's 'soft power' rather than its military strength would define the country's power and greatness. See Putin, 'Russia at the Turn of the Millennium'. Moreover, it was probably the Kursk catastrophe in August 2000 that forced Putin to accept the fact that the nuclear arsenal was a huge problem rather than a usable asset for Russia. See Pavel Baev, 'Putin Reconstitutes Russia's Great Power Status', in *PONARS Policy Memo* (Washington DC: Centre for Strategic and International Studies, November 2003).

101. Putin's views on the development of the mineral sector in Russia were set out in his dissertation, an abstract of which was published. Although the dissertation has since been found to be largely plagiarized and possibly ghostwritten, it is still reflective of Putin's perspectives since he authorized its publication in his name. See Harley Balzer, 'The Putin Thesis and Russian Energy Policy', *Post-Soviet Affairs* 21, no. 3 (2005): 210–225; Martha Brill Olcott, 'Vladimir Putin and the Geopolitics of Oil' (James A. Baker III Institute for Public Policy of Rice University, Texas, October 2004); and David R. Sands, 'Researchers Peg Putin as a Plagiarist over Thesis', *The Washington Times*, 25 March 2006.

102. See Peter Lavelle, 'OPEC Dethroned, Putin's "KremPEC" Arrives', *Untimely Thoughts*, 16 August 2004 (available from http://www.untimely-thoughts.com); Sergey Pokrovskiy, 'Kreml'neftegaz: rasmyshleniya u paradnogo pod'ezda', *Neftegazovaya Vertikal'* (no. 16) 2004; Elena Chinyaeva and Peter Rutland, 'The Yukos Affair: Politics trumps Economics', *Russia & Eurasia Review*, 22 July 2003; and Artem Yeyskov, 'Nationalizatsiya YUKOSa', *Izvestlya*, 3 July 2004.

103. Rail is an alternative means of transporting crude, but is three times more expensive than pipelines. See Mikhail Khodorkovskiy, 'Integration and Consolidation in the Russian Oil Industry', (Presentation at Renaissance Capital's Annual Investors' Conference, Moscow: June 2004). Hence, pipelines remain the dominant mode of transportation, accounting for around 85% of Russian oil exports, as cited by *Russia Oil & Gas Yearbook 2003: Piping Growth* (Moscow: Renaissance Capital, July 2003), p. 75.

104. From Semen Kukes, 'Zven'ya odnoy tsepi', *Neftegazovaya vertikal'*, no. 11, 2003.

105. For instance, in a reference to Transneft' and its monopoly over pipelines, he remarked, 'Don't tell me where to invest my money. I have my own plans and I'm ready to take risks.' Quote cited by Michael S. Lelyveld,

'Russia: Oil Giant, Pipeline Monopoly at Odds', *Radio Free Europe/Radio Liberty*, 29 January 2003 (available from http://www.rferl.org). In contrast, Alekperov of LUKoil was careful not to mention Transneft' by name, preferring instead to note that 'state monopolism in any of its manifestations hinders the development of the Russian oil and gas sector'. From Sergei Blagov, 'Russia's Pipeline Game', *Financial Sense Online*, 15 June 2004 (available at http://www.financialsense.com/editorials/kwr/2004/0615.html).

106. Current Russian legislation provides equal access to all oil companies.

107. From 'YUKOS nameren poluchit' 50% v proekte "Angarsk-Datsin"', *Vremya novostey*, 10 December 2002.

108. These plans were thrown into disarray when Kas'yanov declared, in January 2003, that current and future oil pipelines in Russia would remain the property of the state, although private investment would be welcome. The current prime minister, Mikhail Fradkov, has likewise reiterated that the oil pipeline network 'is our infrastructure and our advantage over other countries, and we must keep an eye on it'. These remarks are from 'Prime Minister: State to Keep Control of Pipelines', *Radio Free Europe/Radio Liberty Business Watch*, 14 January 2003; and Denis Rebrov, 'Fradkov truby ne otdast', *Vremya novostey*, 12 April 2004.

109. Quote by Pavel Kushnir, an oil analyst at United Financial Group, as cited by Catherine Belton, 'Kremlin, Big Oil on Collision Course', *Moscow Times*, 28 January 2003.

110. The Angarsk–Daqing route totals 2400 km, whereas the alternative is 3800–4200 km depending on the actual routing.

111. Following a study on the environmental impact of the Far Eastern pipeline, Transneft' agreed to re-route its proposed pipeline further away from Lake Baykal, with Tayshet (in Irkutsk) and not Angarsk as the starting point (Tat'yana Zykova, 'Nefteprovod "otodvinut" na 400 kilometrov ot Baikala', *Rossiyskaya gazeta*, 24 May 2006). On 30 December 2004, the Russian government finally agreed to proceed with the Tayshet-Perevoznaya (near Nakhodka) pipeline project, as reported in Varvara Aglanish'yan, 'Premer uspel v poslednyy moment: pravitel'stvo razreshilo stroit' nefteprovod', *Izvestiya*, 11 January 2005.

112. Vladislav Shorokhov, 'Truboprovody: rossiysko-amerikanskiy energodialog', *Ekspert*, 7 April 2003. For a similar argument that Khodorkovskiy undermined the state's prerogative in foreign policy issues, see Aleksandr Postnikov, 'Delo "YUKOSa"', *Novoe vremya*, 28 December 2003.

113. Putin, 'Russia at the Turn of the Millennium'.

114. See Paul Merolli, 'Yukos Deal Sounds Right for Exxon, Chevron', *The Oil Daily*, 15 September 2003; and Gregory White and Anita Raghavan, 'Exxon Signals Interest in a Deal with Yukos', *Wall Street Journal Europe*, 12 December 2003. For Russian scholar Harley Balzer, this is the 'most convincing' explanation for the YUKOS affair (see Balzer, 'The Putin Energy Thesis and Russian Foreign Policy').

115. For more information on the basis of the Al'fa group's close association with Putin, see Chapter 5.

116. 'Implications of the Yukos Scandal for Russian Domestic Politics: A Discussion Meeting with Lilia Shevtsova', (Washington DC: Carnegie Endowment for International Peace, 16 September 2004). For a similar

perspective, see Tat'yana Gurova and Aleksandr Privalov, 'Vybor prezidenta: my teryaem ego!', *Ekspert*, 3 November 2003.

117. Quote by an anonymous expert as cited in Maksim Blant, 'Ne YUKOSom edinym', *Yezhenedel'nyy zhurnal*, 15 November 2004. Putin's ambitions to make Russia an energy superpower is also discussed by Fiona Hill, 'Oil, Gas and Russia's Revival' (Report for the Foreign Policy Centre, September 2004).

118. Masha Lipman, 'The Taming of a Tycoon – and of Russia', *The Washington Post*, 16 June 2004.

119. From Tchurilov, *Lifeblood of Empire*, p. 224.

120. Recounted in author's interview with Mr Viktor Ivanenko.

121. See William Tompson, 'Putin and the "Oligarchs": A Two-Sided Commitment Problem', in *Leading Russia: Putin in Perspective*, ed. Alex Pravda (Oxford: Oxford University Press, 2005).

122. According to Khodorkovskiy's security chief, in the event of the tycoon being arrested, 'the calculation was that the West would talk to Putin and ask him why he was holding such people behind bars . . . The calculation was that Putin would receive a friendly warning that this was just not done by people who sit on the G8'. As cited in Catherine Belton, 'The Friends and Foes of a Rising Oilman', *Moscow Times*, 27 May 2005.

123. Gevorkyan, Timakova and Kolesnikov, *First Person*, p. 186.

124. Author's interview with Mr Sergey Molozhaviy.

125. The roles of these and other bureaucracies pertaining to privatization policy are from *The Investment Environment in the Russian Federation: Laws, Policies and Institutions* (Paris: Organisation for Economic Co-operation and Development, 2001), p. 149; and Nelson and Kuzes, *Property to the People*, p. 126.

126. From author's interview with Mr Sergey Generalov.

127. Author's interview with Dr Rair Simonyan (Moscow).

128. Only Chubays and Igor' Yurgens (chairman of the RSPP) publicly voiced their disapproval about Khodorkovskiy's arrest. See Anzhela Sikamova, Ol'ga Tropkina, Yevgeniy Mazin and Petr Orekhin, 'Oligarkhi na pereput'e: soratniki ne toropyatsya podderzat' Mikhaila Khodorkovskogo', *Nezavisimaya gazeta*, 28 October 2003.

129. As quoted by Julia Latynina, 'Garden Parties and the Criminal Code', *St Petersburg Times*, 30 September 2003.

130. As noted in various entries in Boris Yeltsin, *Against the Grain: An Autobiography*, trans. Michael Glenny (New York: Summit Books, 1990). Alexander Lukin has also noted that for Yel'tsin, democracy was not always an instrument to consolidate power and has cited the case of Yel'tsin pursuing bilateral treaties with Russia's republics even after he had won election as the country's president. See Lukin, *The Political Culture of the Russian 'Democrats'*, p. 293.

131. The book does not share Olcott's tendency towards an overly deterministic explanation of the YUKOS affair, that is, that Putin was merely following up on ideas espoused in his dissertation about the role of the state in developing mineral resources (see Olcott, 'Vladimir Putin and the Geopolitics of Oil'). Such a perspective fails to satisfactorily explain why the informal July 2000 agreement was created in the first place, or why Putin agreed to proceed with the sale of Slavneft' in 2002.

4 Case Study of Slavneft'

1. Roland Nash, chief economist at a leading investment bank, Renaissance Capital, as cited by Gregory Feifer, 'The Kremlin's Big Sell-Off', *Moscow Times*, 19 December 2000.
2. Vladimir Malin, head of the Russian Federal Property Fund, as cited in 'Russian Government sells Onako state oil company for $1.08 billion', *The Associated Press*, 19 September 2000.
3. Farit Gazizullin, Minister of Property Relations (the renamed GKI), as cited in Ibid.
4. Boris Kagarlitsky, '"Political Capitalism" and Corruption in Russia', *Links* (May–August 2002) available at http://www.dsp.org.au/links/back/issue21/Kagarlitsky.htm.
5. Author's interview with Mr Aleksandr Putilov.
6. See Leonid Brodskiy, 'Novye naznacheneniya v pravitel'stve: Anatoliy Fomin – pervyy zamministra topliva i energetiki', *Kommersant'*, 8 April 1993.
7. See *Energy Policies of the Russian Federation*, p. 97.
8. Author's interview with Mr Anatoliy Fomin, ex-deputy Minister of Fuel & Energy of the Russian Federation and former President of Slavneft', 9 February 2004, Moscow.
9. 'Zastrelen neftyanoy magnat', *Kommersant'*, 6 August 1994.
10. See Vadim Belykh, 'Chernaya krov'', *Izvestiya*, 4 January 1997. Fomin still retains a close relationship with Alafinov. Fomin is today the president of Bank Yugra while Alafinov is a member of the Board of Directors. Both of them, along with some of their family members, are the sole shareholders of the bank.
11. Arthur F. Bentley, *The Process of Government: A Study of Social Pressures*, 1995 ed. (New Brunswick: Transaction Publishers), p. 117.
12. Kenneth A. Shepsle, 'Comment on Derthick and Quirk', in *Regulatory Policy and the Social Sciences*, ed. Roger G. Noll (Berkeley: University of California Press, 1985), p. 233.
13. See, for example, Robert C. Lieberman, 'Ideas, Institutions and Political Order: Explaining Political Change', *American Political Science Review* 96, no. 4 (December 2002); John Kurt Jacobsen, 'Much Ado about Ideas: The Cognitive Factor in Economic Policy', *World Politics* 47, no. 2 (1995): 283–310; and Dietmar Braun, 'Interests or Ideas?', in *Public Policy and Political Ideas*, ed. Dietmar Braun and Andreas Busch (Cheltenham: Edward Elgar, 1999).
14. See Igor V. Timofeyev, 'The Development of Russian Liberal Thought since 1985', in *The Demise of Marxism–Leninism in Russia*, ed. Archie Brown (Basingstoke: Palgrave Macmillan, 2004); and Lukin, *The Political Culture of the Russian 'Democrats'*.
15. See Barry R. Weingast, 'A Rational Choice Perspective on the Role of Ideas: Shared Belief Systems and State Sovereignty in International Co-operation', *Politics & Society* 23, no. 4 (1995): 449–464; Geoffrey Garrett and Barry R. Weingast, 'Ideas, Interests and Institutions: Constructing the European Community's Internal Market,' in *Ideas and Foreign Policy: Beliefs, Institutions and Political Change*, ed. Judith Goldstein and Robert O. Keohane (Ithaca: Cornell University Press, 1993); James G. March and Johan P. Olsen, 'The New Institutionalism: Organisational Factors in Political Life', *American Political*

Science Review 78, no. 3 (1984): 738–749; and Paul A. Sabatier and Hank C. Jenkins-Smith, 'The Advocacy Coalition Framework: An Assessment', in *Theories of the Policy Process*, ed. Paul A. Sabatier (Boulder, Colorado: Westview Press, 1998).

16. The term is borrowed from Margot Light, 'Foreign Policy Thinking', in *Internal Factors in Russian Foreign Policy*, ed. Neil Malcolm et al. (Oxford: Clarendon Press, 1996), p. 34. See also Adranik Migranyan, 'Russia and the Near Abroad: The Entire Space of the Former USSR is a Sphere of Russia's Vital Interests', *Nezavisimaya gazeta*, 18 January 1994 (compiled in *Current Digest of the Post-Soviet Press*, XLVI, 7, 1994, p. 6).

17. Russia's defence of its various interests in the CIS is analysed by Alex Pravda, 'Foreign Policy', in *Developments in Russian Politics 5*, ed. Stephen White, Alex Pravda and Zvi Gitelman (Basingstoke: Palgrave Macmillan, 2001); Dov Lynch, *Russian Peacekeeping Strategies in the CIS: The Cases of Moldova, Georgia and Tajikistan* (Basingstoke: Palgrave Macmillan, 2000); and Thomas Goltz, 'Letter from Eurasia: The Hidden Russian Hand', *Foreign Policy*, no. 92 (1993): 92–116.

18. Aleksandr Romanikhin, 'Neftyanye kompanii slishkom uvleklis' eksportom syroy nefti: Rossiya zakupaet vse bol'she benzina na granitsey', *Segodnya*, 14 August 1996.

19. See 'Russia Prepares to Take a Hard Line with the CIS', *International Herald Tribune*, 20–21 October 1994; and 'Russian Subsidies for Belarus', *Financial Times Energy Newsletters: East European Energy Report*, 26 August 1994.

20. See 'Belarussian Supply Company becomes Part of Gazprom', *BBC Summary of World Broadcasts*, 17 September 1993; and Mikhail Shimansky, 'Opposition Accuses Government of Selling off Belarus', *Izvestiya*, 15 December 1992 (compiled in the *Current Digest of the Post-Soviet Press*, XLV, 50, 1994, p. 23).

21. From author's interviews with Mr Anatoliy Fomin and Mr Dmitriy Romanov.

22. As quoted by Lee Hockstader, 'Belarus Votes to Go Back to the Future: Winner reflects Nostalgia for good old Soviet days', *The Washington Post*, 16 July 1994. The deputy prime minister appointed by Lukashenko was Valeriy Kokarev.

23. See Romanikhin, 'Neftyanye kompanii slishkom uvleklis' eksportom syroy nefti'.

24. See Anna Baneva, 'Rossiyskie neftyanye kompanii v Belorussii: Rossiyskaya neft' obespechit rekonstruktsiyu belorusskikh zavodov', *Kommersant'-Daily*, 23 November 1995; and 'Belarus to Swap Slavneft Stake for Refinery Stake', *Financial Times Energy Newsletters: East European Energy Report*, 1 April 1996.

25. Author's interview with Mr Dmitriy Romanov.

26. It was widely assumed that Kuz'min's death was a result of business disputes within the oil industry. See the report in the previously cited article by Belykh, "Chernaya krov" .

27. For examples of such optimistic assessments, see Theodore Kim, 'Chubais Forces the Pace', *Euromoney*, September 1997; and Chrystia Freeland, 'Russia: New Signs of Hope Appear', *Financial Times (London)*, 19 September 1997.

28. Author's interview with Mr Dmitriy Romanov.

29. 'Uderzhit li Duma 'Slavneft"?', *Rossiyskaya gazeta*, 18 September 1998.

30. Al'fa issued a statement denying its involvement in the law suit. See 'Konkurs po 'Slavnefti" sorval monter Samodurov', *Russkiy Telegraf*, 24 March 1998;

and Aleksandr Vladimirov, 'Konkurs po prodazhe aktsiy ne sostoyalsya', *Nezavisimaya gazeta*, 28 March 1998.

31. From author's interview with Mr Dmitriy Romanov.
32. See Sujata Rao, 'Tyumen Oil Auction Suspended by Court', *Moscow Times*, 21 November 1997.
33. See, for example, Gary Peach, 'Big Oil Sells-Offs Laudable, But Haste May Make Waste', *Moscow Times*, 11 November 1997.
34. Fomin's formidable reputation was a source of consternation among GKI officials. According to its deputy minister at that time, no one wanted to put their name on the order to dismiss him as president of Slavneft' in 1998: 'everybody in the ministry was afraid to do it. They were afraid that they could be killed for that because he had a very bad image . . . Fomin is quite an unusual man, an extraordinary man'. As recounted in author's interview with Mr Sergey Molozhaviy, former deputy minister of GKI, 10 February 2004, Moscow.
35. This observation refers only to auctions for the first share placement of a company. Subsequent placements of additional shares, such as those by TNK, are excluded.
36. Figure cited in *Russian Economy: Trends and Perspectives* (Moscow: Institute for Economy in Transition, January 2003).
37. This new decree 1148, dated 4 November 1997, replaced the corresponding provision contained in the previous decree 1403 of November 1992. It declared that the removal of the limit would apply to auctions of oil companies announced after the introduction of the decree (for example, Slavneft' and Rosneft') and not to those announced prior to the decree (such as KomiTEK, Norsi Oil, VNK).
38. Author's interview with Mr Dmitriy Romanov.
39. Jeanne Whalen, 'Wary Bidders Dampen "98 Oil Sell-Offs"', *Moscow Times*, 24 January 1998. Similar claims are cited in 'Russia Struggling to Sell-Off Oil Jewels in Crown', *Agence France Press*, 13 May 1998; and 'World Market Instability Threatens Russian Privatisation', *Agence France Press*, 11 November 1997.
40. Figure cited in *Russian Economy in 1998: Trends and Perspectives* (Moscow: Institute for Economy in Transition, 1999).
41. For example, the market capitalization of YUKOS fell from US$9.3 billion to US$1.4 billion to US$0.3 billion in 1997, 1998 and 1999, respectively, while corresponding figures for Sibneft' were US$5.4 billion, US$0.9 billion and US$1.5 billion. Figures from annual company ratings compiled by *Ekspert* magazine are available at http://www.expert.ru/expert/ratings/exp200.
42. Prices are based on the average annual Urals spot crude prices, as compiled in *Energy Prices & Taxes: 3rd Quarter 2004* (Paris: International Energy Agency, 2004), p. 4.
43. Officially, Fomin was dismissed because of mounting debts owed by Mozyr to Slavneft' after April 1998. For an alternative perspective that links Fomin's dismissal to his conflicts with Sergey Kiriyenko, see Vladimir Vostrukhin and Pavl Kuybyshev, 'Rossiyskie kompanii proigrali frantsuzskoy v bor'be za neftyanoy rynok', *Kommersant'-Daily*, 2 October 1998. According to a former vice-president of Slavneft', Duma used to remark to him that what was significant was not that he had been appointed but, rather, that Fomin had

been dismissed; in this connection, it was opined during the interview that the 'young reformers' could have appointed 'any Ivanov, Petrov or Sidorov who happened to be in the right place' at that time (from author's interview with Dmitriy Romanov).

44. Cited in 'Slavneft President Holds that Government Should Possess Controlling Shares of Oil Company', *SKRIN Market and Corporate News*, 22 October 1999. See also 'Duma Throws in the Towel at Slavneft', *Nefte Compass*, 11 November 1999.

45. Petr Sapozhnikov, 'Duma pod voprosom', *Kommersant'-Daily*, 11 November 1999.

46. For these and more on Gutseriev's views, see Lyudmila Romanova, 'Mikhail Gutseriev: Prezident "Slavnefti"' – dolzhnost' skoree politicheskaya', *Nezavisimaya gazeta*, 12 February 2000; Valeria Korchargina, 'New Boss has Ambitious Vision for Slavneft', *Moscow Times*, 8 February 2000; and Petr Sapozhnikov, '"Slavneft"' mozhet vernut' TNK gosudarstvu', *Kommersant'-Daily*, 21 July 2000.

47. The announcement was made in late May 2000 by the head of TNK, Simon Kukes, who also revealed that TNK would participate in the auctions for 19.68% of Slavneft'.

48. 'Boy bez pravil: ili kto budet kontrolirovat' "Slavneft"?,' *Neft' i Kapital*, no. 9, 2000.

49. Gutseriev denied any involvement in the raids and threatened to sue TNK for slander. See Mikhail Zimin, 'Malen'kaya neftyanaya voyna: Iskhod konflikta mezhdu TNK i "Slavneft"'yu' poka ne reshen', *Segodnya*, 5 August 2000 and the interview with Gutseriev by Sapozhnikov, '"Slavneft"' "mozhet vernut" TNK gosudarstvu'.

50. Gutseriev's relationship with Rushailo is documented in various sources including author's interview with Mr Andrey Shtorkh, former vice-president of Slavneft', 1 October 2003; Yuliya Latynina, 'Ukreplenie gorizontali vlasti', *Novaya gazeta*, 29 April 2002; and Yuliya Latynina, 'Oligarkh, kotoryy ne sel chizhika', *Novaya gazeta*, 20 May 2002.

51. For a study of the different types of private protection arrangements in Russian business, see Vadim Volkov, *Violent Entrepreneurs: The Use of Force in the Making of Russian Capitalism* (Ithaca: Cornell University Press, 2002); and Frederico Varese, *The Russian Mafia: Private Protection in a New Market Economy* (Oxford: Oxford University Press, 2001).

52. See interview by Sapozhnikov, '"Slavneft"' mozhet vernut' TNK gosudarstvu'.

53. In fact, according to the 1997 privatization law, the Duma had to right to approve lists of strategic enterprises for privatization. However, its refusal to approve the lists since 1997 had always been overcome by the issuing of presidential decrees on the matter. For the Property Ministry, this represented an undesirable legal loophole and it wanted to permanently deny this channel to the Duma by introducing a new, revised law on privatization.

54. Sergey Pravosudov, '"Slavneft' i TNK vedut pozitsionnuyu voynu', *NG Politekonomia*, 20 June 2000.

55. Freeland, *Sale of the Century*, pp. 102–103.

56. The charges were later dropped after Sukhanov was elected president of Slavneft'. For an account of the investigations into Sukhanov's role, see

Aleksandr Petrovykh, 'Skandaly: "Slavneft"'yu' zanimayutsya v Gosdume, MVD i Genprokurature', *Novaya gazeta*, 6 May 2002, Anna Raff, 'Sukhanov Elected Slavneft President', *Moscow Times*, 14 May 2002; and Torrey Clark, 'Police Probe Slavneft Execs', *Moscow Times*, 13 May 2002.

57. The principal–agent framework has been used to analyse relations between the legislature and bureaucracy. See, for instance, Matthew McCubbins, Roger G. Noll and Barry R. Weingast, 'Structure and Process, Politics and Policy: Administrative Arrangements and the Political Control of Agencies', *Virginia Law Review* 75 (1989): 431–483. For an analysis of the principal–agent problem in general and its solution, see Kenneth A. Shepsle and Mark S. Bonchek, *Analysing Politics: Rationality, Behaviour and Institutions* (New York: W W Norton & Company, 1997), pp. 360–279; and Jean-Jacques Laffont and David Martimort, *The Theory of Incentives: The Principal–Agent Model* (Princeton: Princeton University Press, 2002).

58. As cited by Ekaterina Drankina, 'Zakat Gutserieva', *Ekspert*, 15 April 2002; and 'Sibneft/TNK Bury the Hatchet over Slavneft', *FSU Energy*, 10 November 2000. Gutseriev, however, denied any inappropriate relationship between BIN bank and Slavneft'. See, for instance, Maksim Rybchenko, 'My nikogda ne byli ch'im-to opornym bankom', *Ekspert*, 22 April 2002.

59. For allegations, see Drankina, 'Zakat Gutserieva'; and 'Slavneft and Megionneftegaz: Privatisation Implications', (Moscow: Renaissance Capital, 22 October 2002), pp. 12–13.

60. The settlement between Gutseriev and Sibneft'–TNK over Varyeganneft' is described in Ol'ga Shevel, 'Chelovek masshtaba', *Sekret firmy*, 21 November 2005 (available at http://www.sf-online.ru); 'Morning Monitor', (Moscow: Renaissance Capital, 14 November 2002); *Russian Market Daily* (Moscow: CentreInvest Group, 14 November 2002); 'Whole Hog: Gutseriev Pursues Vertical Integration', *Nefte Compass*, 9 December 2002; and 'Majors Line Up for Slavneft Sale of the Year', *Nefte Compass*, 20 November 2002.

61. Gutseriev was forced to sell Russneft' to an intermediary supposedly acting on behalf of unnamed state officials and is a tax fugitive from the Russian authorities. For background on the rise and fall of Russneft', see Greg Walters, 'A New Oil Major From Nowhere', *Moscow Times*, 3 August 2005; 'Russneft's Arrival', *EIU Viewswire*, 10 February 2006; Yevlaliya Samedova and Irina Reznik, 'Rossiya pomenyala neftyanykh igrokov v slovakii', *Nezavisimaya gazeta*, 8 February 2006; 'Kremlin Turns Guns on Independent Oil Producer', *Nefte Compass*, 1 February 2007; Yuliya Latynina, 'A teper' "Russneft"'?', *Novaya gazeta*, 17 May 2007; and Natal'ya Grib, Alek Akhundov, Denis Rebrov and Ol'ga Pleshanova, 'Mikhailu Gutserievu naznachili vykhodnoe posobie', *Kommersant'-Daily*, 30 July 2007.

62. As cited by Sapozhnikov, '"Slavneft"' mozhet vernut' TNK gosudarstvu'.

63. As cited by Oleg Chernitskiy, 'Privatizatsii: "Slavnefti"' ne budet', *Vremya Novostey*, 21 December 2000.

64. The Gutseriev–Rushailo relationship was also reflected in the hostage situation in Chechnya. Using his high standing in the North Caucasus, Gutseriev would negotiate for the freedom of some hostages and turn them over to Rushailo, who then took credit for his Ministry's effectiveness in freeing hostages in Chechnya. From author's interview with Mr Andrey Shtorkh; and Shevel, 'Chelovek masshtaba'.

65. Julia Latynina, 'Perilous Incompetence in Ingushetia', *Moscow Times*, 10 April 2002. Apart from federal subsidies, the Russian government also paid out large sums for the release of Russians taken hostage by the Chechens. Most of the mediation was done by the Interior Ministry and the Gutseriev clan, leading to speculation that some of the money may have been diverted. Gutseriev's role in release of hostages was described by one of his former colleagues, as noted in author's interview with Mr Andrey Shtorkh.

66. For an account of the origins and activities of the BIN group in Ingushetia, see Mark Whitehouse, 'From War Zone to Tax Haven', *Moscow Times*, 4 March 1997.

67. See Romanova, 'Mikhail Gutseriev: Prezident "Slavnefti"' – dolzhnost' skoree politicheskaya'. In the same interview, Gutseriev denied that his position at Slavneft' was the result of 'political trading between Ruslan Aushev and Vladimir Putin'.

68. Quote by Danil Khachaturov, a former spokesman for Gutseriev at Slavneft', as cited by Yuliya Bushueva and Elena Berezanskaya, 'Mezhpromslavneft', *Vedomosti*, 17 April 2002.

69. From 'Brat otvetil za brata', *Kommersant'-Daily*, 22 April 2002; and from author's interview with Mr Andrey Shtorkh. According to the latter, Abramovich repeatedly warned Gutseriev not to get involved in the Ingush elections as it could cost him his position at Slavneft'.

70. See Vladimir Yanchenkov, 'Dlya izbraniya prezidenta Ingushetii trebuetsya vtoroy tur', *Trud*, 9 April 2002. Slavneft' denied allegations that the company had financed Khamzat Gutseriev's campaign. See 'Russian Oil Company Denies Involvement in Ingush Election Campaign', *BBC Monitoring: Former Soviet Union*, 5 April 2002.

71. See Liz Fuller, 'Ingushetia's President Bows Out', *Radio Free Europe/Radio Liberty Caucasus Report*, 3 January 2002.

72. From Peter Rutland, 'Putin's Path to Power', *Post-Soviet Affairs* 16, no. 4 (2000), p. 334.

73. For an in-depth analysis of Putin's early reforms, see Archie Brown, 'Vladimir Putin and the Reaffirmation of Central State Power', *Post-Soviet Affairs* 17, no. 1 (2001): 45–55; and Gordon M. Hahn, 'The Impact of Putin's Federative Reforms on Democratisation in Russia', *Post-Soviet Affairs* 19, no. 2 (2003): 114–153.

74. Julia Latynina, 'Make an Honest Woman of Slavneft', *Moscow Times*, 11 December 2002.

75. 'TNK and Sibneft Join Forces to Defend Slavneft', *FSU Energy*, 26 April 2002.

76. The corporate battle at Slavneft' received a lot of coverage in the Russian newspapers. A useful summary with suggestions for further reading may be found in 'Slavneft Headquarters Reopens amid Questions about Russian President's Leadership', *Radio Free Europe/Radio Liberty Business Watch*, 9 July 2002.

77. This writ was later annulled by a ruling from a court in Volgograd which upheld the legality of the said meeting.

78. A profile of Pugachev and his alleged links to Putin and the *siloviki* may be found in Il'ya Bulavinov, 'Politicheskiy rasklad: koloda Rossiyskoy Federatsii', *Kommersant'-Vlast'*, 1 December 2003. Pugachev subsequently attempted to sue an analyst for making such allegations, as reported in Vitaliy Ivanov, 'Pugachev trebuet $1mln', *Vedomosti*, 10 October 2003.

79. See, for instance, 'Sergey Bogdanchikov: Turning Rosneft around', *Radio Free Europe/Radio Liberty Business Watch*, 9 July 2002; and Clark, 'Police Probe Slavneft Execs'. A Slavneft' insider, however, expressed reservations about whether Bogdanchikov had actually sanctioned Baranovskiy's actions at Slavneft', despite the fact that Baranovskiy was also the vice-president at Rosneft' (from author's interview with Mr Andrey Shtorkh).

80. Aleksey Makarkin and Sergey Savushkin, 'Rol' lichhosti v istorii neftyanoy goskompanii', *Neft' i Kapital*, no. 10, 2002.

81. For details, see Yuliya Bushueva and Aleksandr Tutushkin, 'Vse-taki part-nery', *Vedomosti*, 15 October 2002; and Isabel Gorst, 'Sibneft and Tyumen Oil Reach Swap Agreement', *Platt's Oilgram News*, 16 October 2002.

82. Andrei Litvinov, 'Abramovich to Add Slavneft to his Empire', *Gazeta.ru*, 11 October 2002 (printed in *Johnson's Russia's List* no. 6488).

83. See Yakov Pappe, *Oligarkhi: Ekonomicheskaya Khronika 1992–2000* (Moscow: Gosudarstvennyy universitet vysshaya shkola ekonomiki, 2000), p. 227; and Nikolay Petrov, 'Gruppa "Al'fa" zakhvatyvaet Dumu', *Novaya gazeta*, 21 February 2000.

84. From author's interview with Mr Sergey Molozhaviy.

85. Irina Reznik, Yuliya Bushueva and Yelizaveta Osetinskaya, '"Slavneft"' odnim lotom', *Vedomosti*, 9 October 2002. The 'two presidents' in the quotation refer to Gutseriev and Sukhanov, each of whom were struggling to become the head of Slavneft' in mid-2002.

86. See, for instance, Boyko, Shleifer and Vishny, *Privatising Russia*; Kokh, *The Selling of the Soviet Empire*; and Andrei Shleifer and Robert W. Vishny, 'Politicians and Firms', *The Quarterly Journal of Economics* 109, no. 4 (1994): 995–1025.

87. There appears to be a consensus in the empirical literature that as a general rule, privately owned firms consistently outperform state-owned enterprises, and that among privately owned firms, differences in ownership types account for the varying degrees of success. See, for instance, William L. Megginson, Robert C. Nash and Matthias von Randenborgh, 'The Financial and Operating Performance of Newly Privatised Firms: An International Empirical Analysis', *Journal of Finance* XLIX (1994): 403–452; Roman Frydman, Cheryl Gray, Marek Hessel and Andrzej Rapaczynski, 'When Does Privatisation Work? The Impact of Private Ownership on Corporate Performance in the Transition Economies' (Stockholm: Stockholm School of Economics, SITE Working Paper no. 142, January 1999); and Simeon Djankov and Peter Murrell, *The Determinants of Enterprise Restructuring in Transition: An Assessment of the Evidence* (Washington DC: World Bank, September 2000). For a dissenting view on the subject, see Alec Nove, *Efficiency Criteria for Nationalised Industries* (London: George Allen & Unwin, 1973). He has argued that nationalized industries such as the railways or electricity should not be judged on commercial criteria, but instead on the intangible benefits they provide to the wider public.

88. Reznik et al., '"Slavneft"' odnim lotom'. Similar sentiments were echoed by Litvinov, 'Abramovich to Add Slavneft to his Oil Empire'; 'Slavneft Sale: Implications and Interpretations', *Radio Free Europe/Radio Liberty Business Watch*, 22 October 2002; and 'Oil Test for Putin's regime', *Euromoney*, December 2002.

89. See 'Slavneft Sale a Safe Bet against Oil Instability', *Moscow Times*, 14 October 2002.
90. Author's interview with Mr Sergey Molozhaviy.
91. Sergey Fedoktin, 'Vse na vybory!', *Rossiyskie vesti*, 16 October 2002.
92. Ibid.
93. Mariya Ignatova, 'Roman s neft'yu: 'Slavyanskaya' neft' stala sibirskoy', *Izvestiya*, 19 December 2002.
94. See Simon, 'Organisation and Markets'.

5 Case Study of Rosneft'

1. Tchurilov, *Lifeblood of Empire*, p. 215.
2. 'Blueprint for Reform: New Policies and Structures for the Russian Oil Industry', (New York: Petroleum Intelligence Weekly, 1992), p. 73. The consultants did issue a disclaimer that the information upon which this report was based was not independently verified and was provided solely by Rosneftegaz.
3. This negative image was rather undeserved since Rosneftegaz proved to be a strong advocate of significant changes to the oil industry, including an increase in the price of oil, the formation of VIOCs and the need for foreign investment. See 'Soviet Flow Skids to Less than 10 Million b/d', *Oil & Gas Journal*, 30 December 1991; and Leyla Boulton, 'Oil Industry Chief Warns of Need for Big Price Increase', *Financial Times (London)*, 17 August 1991.
4. John Lloyd, 'Russian Oil Shake-Up Raises Foreign Hopes', *Financial Times (London)*, 19 February 1993. Churilov, however, denied Lloyd's claim that there was a high level of tension between Mintopenergo and Rosneftegaz (from author's interview with Mr Lev Churilov).
5. Between 1987 and 1990, Shafranik was general manager of the oil producer, Langepas, while Putilov was his counterpart at Uray.
6. Stepan Avdeev, 'Preobranzovanie neftyanogo kompleksa: Neft' vzyata pod goskontrol' novym sposobom', *Kommersant'*, 14 November 1992; and 'Srednee zveno: gosudarstvennoe predpriyatnie "Rosneft'"', *Ekonomika i zhizn'*, no. 35, August 1993.
7. The quote by Putilov is from 'Rosneft Head Sees New Plan Soon to Reorganise Russian Oil Companies', *Platt's Oilgram News*, 2 June 1994. Putilov also submitted a blueprint for Rosneft' to retain seven oil producers and four refineries, as described in 'Reforms Divide Russian Oil Giants', *Lloyd's List*, 23 December 1994.
8. See the account by Leonid Berres, 'Konflikt Goskomimushchestva s "Rosneft'yu": aktsionery "Purneftegaza" ne soglasilis' s pravitel'stvom', *Kommersant'*, 27 July 1994.
9. See, for example, 'Russian National Oil Company Plans Criticised by Sidanko', *Financial Times Energy Newsletters: East European Energy Report*, 27 January 1995; and Aleksei Sukhodoev, 'Funktsii rossiyskoy natsional'noy neftyanoy kompanii', *Kommersant'*, 4 August 1994.
10. As quoted in 'Reforms Divide Russian Oil Giants'.
11. Quote by Dmitriy Romanov as cited by Mikhail Dubik, 'Officials Predict an End to Soviet-Era Rosneft', *Moscow Times*, 7 June 1994. A report commissioned jointly by the World Bank and Mintopenergo held similar views.

See Robert Corzine, 'State Oil Company "Wrong" for Russia', *Financial Times (London)*, 1 March 1995. For reactions by leading Russian officials and businessmen, see Andrey Konoplyanik, *Rossiya na formiruyushchemsya evroaziatskom energeticheskom prostranstve: problemy konkurentosposobnosti* (Moscow: Nestor, 2004), pp. 454–464.

12. The plan was not formally approved.

13. Putilov was, in fact, so confident that Chernomyrdin would keep his word regarding the structure of Rosneft' that he left Russia soon after for his summer vacation in France, as recounted in author's interview with Dr Rair Simonyan (Moscow).

14. According to presidential decree 1403 of November 1992, the stakes owned by foreigners in an oil company were limited to 15% of its authorized capital. The removal of this limit was aimed at attracting foreign interest and was not retroactive, that is, it was only applicable to privatizations that were to be announced henceforth. See Natal'ya Samoylova and Aleksandr Malyutin, 'Kreml' gotov prodavat' "Rosneft" inostrantsam', *Kommersant'-Daily*, 4 November 1997; and Natal'ya Samoylova, 'Prezident razreshil inostrantsam pokupat' neftyanye kompanii', *Kommersant'-Daily*, 6 November 1997.

15. Aleksandr Budberg, 'BAB prokhodit kak khozyain . . .' *Moskovskiy komsomolets*, 28 November 1997.

16. As cited by Iain Esau, 'Shell Pulls Out of Rosneft Bid Consortium', *Lloyd's List*, 4 July 1998. A similar reason was given by British Petroleum/ONEKSIMbank. See Jeanne Whalen, 'Unexim Drops Out of Rosneft Auction', *Moscow Times*, 7 July 1998.

17. Author's interview with Dr Rair Simonyan (Moscow).

18. Ibid. Berezovskiy has also used the strategy of 'privatising management' to his control over other companies, such as Aeroflot, ORT and the constituents of Sibneft'.

19. Leonid Proshin, 'Aleksandru Putilovu nashli mesto v "Rosnefti"', *Kommersant'-Daily*, 7 May 1997. As part of the campaign to reinstate Putilov, Simonyan and his colleagues also publicized the matter in the press and had meetings with Nemtsov.

20. Pre-tax profits for the first three quarters of 1997 had declined by 19.4% over the same period in 1996, while refining volumes had fallen by 11% despite a rise in oil-production levels. Putilov claimed that these figures indicated that under Bespalov, some of the crude produced by Rosneft' was being sold at a low price to Omsk refinery (owned by Berezovskiy's Sibneft'), which was charging high prices to refine the oil. See 'Rosneft Privatisation Heads for Final Round', *Financial Times Energy Newsletters: East European Energy Report*, 1 December 1997; and Gary Peach, 'Raid of the Rosneft Cookie Jar May Just Leave Crumbs', *Moscow Times*, 12 May 1998.

21. For details, see Natal'ya Gotova, 'Killer Berezovskiy metit v Nemtsova', *Moskovskiy komsomolets*, 29 November 1997.

22. For more information, see Budberg, 'BAB prokhodit kak khozyain . . .'; and Yuliya Panfilova, 'Delezh "Rosnefti" nachalsya iznutri', *Kommersant'-Daily*, 24 October 1997.

23. See 'Registrator "Rosnefti": ni nashim, ni vashim', *Kommersant'-Daily*, 6 December 1997; and Yuliya Panfilova, 'Reestr aktsionerov "Rosnefti" budet vesti kompaniya Borisa Berezovskogo', *Kommersant'-Daily*, 15 October 1997.

A registrar is required by law to maintain information on the distribution of a company's shareholdings. In Russia, registrars affiliated to a company's management sometimes fail to record share purchases by 'hostile' investors or they allow shares to be transferred to shell companies owned by management. See Frederico Varese, *The Russian Mafia: Private Protection in a New Market Economy* (Oxford: Oxford University Press, 2001), pp. 32–33.

24. Bespalov's relationship with Berezovskiy is also mentioned by Andrey Belyaev, 'Viktoru Chernomyrdinu razvyazali ruki', *Kommersant'-Daily*, 27 November 1997; Jane Upperton, 'Management Face-Off Nears at Rosneft', *Platt's Oilgram News*, 25 July 1997; and Gotova, 'Killer Berezovskiy metit v Nemtsova'. Berezovskiy was engaged in advanced negotiations with Gazprom at this time and had named Gazprom's president, Rem Vyakhirev, as chairman of Sibneft' in an attempt to seal the alliance. However, Gazprom abruptly rejected Berezovskiy in favour of LUKoil and Royal Dutch/Shell.

25. At that time, the government claimed that the compromise was in line with new corporate guidelines that came into effect in June 1996 that prohibited the posts of chairman and president from being held by the same person simultaneously. This is, however, a weak justification since candidates other than the protagonists could have been considered for the posts.

26. See Tat'yana Koshkareva and Rustam Narzikulov, 'V preddverii auktsiona po "Rosnefti"', *Nezavisimaya gazeta*, 21 February 1998. Aleksandr Kazakov was chairman of Gazprom from mid-1996 to 1998. He was previously the head of GKI and deputy head of Yel'tsin's presidential administration.

27. Author's interview with Mr Dmitriy Romanov.

28. Tat'yana Koshkareva and Rustam Narzikulov, 'Liberaly delyat neftyanuyu sobstvennost'', *Nezavisimaya gazeta*, 10 December 1998. A similar conclusion was reached by Celestine Bohlen, 'Energy Companies are Tools in Russian Power Struggle', *The New York Times*, 11 October 1999.

29. Vyacheslav Nikonov, 'Vtoraya liberal'naya revolyutsiya', *Nezavisimaya gazeta*, 10 April 1997.

30. 'New Energy Minister Blasts Delay in Rosneft Privatisation', *BBC Summary of World Broadcasts*, 5 December 1997.

31. Natal'ya Gotova, 'Semerykh odnim udarom', *Moskovskiy komsomolets*, 23 December 1997. Omsk was, in fact, the only refinery belonging to Sibneft'. Chubays, who was head of the Emergency Committee for Taxes and Budget, also ordered the arrest and disposal of assets belonging to SIDANKO's refinery for back taxes amounting to 766 billion rubles. Given that the owner of SIDANKO, Vladimir Potanin, was closely allied with Chubays, this was interpreted as the latter's attempt to appear to be even-handed in his relationship with the tycoons, particularly, since there was idle refining capacity at SIDANKO. In the end, both refineries paid off their debts without any forced liquidation of assets.

32. Quote by Aleksandr Bekker, 'Who Will Take on the Redhead?', *Moskovskiye novosti*, 21 December 1997 (compiled in *Current Digest of the Post-Soviet Press*, XLIX, no. 51, 1997, pp. 6–7).

33. Chrystia Freeland, 'New Hitch in Rosneft Sale', *Financial Times (London)*, 30 January 1998. Likewise, Kiriyenko explained that the 75%-plus-one proposal was forwarded in December 1997 because it was strategic investors with a long-term interest that were most interested in bidding for a stake in Rosneft',

rather than portfolio investors who have shorter time horizons and expect a quick return. See 'Rosneft Privatisation Heads for Final Round'. Earlier privatization plans had called for the sale of smaller stakes, as reported in 'Utverzhden plan privatisatsii "Rosnefti"'; and Maksim Kashulinskiy, 'Bor'ba za "Rosneft" obeshchaet byt' zharkoy', *Segodnya*, 30 July 1997.

34. The sentiment that Rosneft' was overvalued was shared by potential bidders, the board of directors of Rosneft' and independent analysts. For a selection of reactions to the asking price, see 'Rosneft Terms to Remain despite Price Complaints', *Financial Times Energy Newsletters: East European Energy Report*, 1 April 1998.

35. As reported by Sujata Rao, 'Rosneft May be Split into Small Stakes', *Moscow Times*, 24 April 1998.

36. As if to confirm Yel'tsin's perceptions about Chernomyrdin, a Russian newspaper wrote that Chernomyrdin's foreign visits and weekly television address at the start of 1998 gave the impression that he was 'more than simply a prime minister', and that the ailing president himself was 'almost unnecessary'. See Elena Dikun and Anatoliy Kostyukov, 'Upotreblenie vlasti', *Obshchaya gazeta*, 9 July 1998.

37. As quoted in 'Kiriyenko Links Shell Putt-Out from Oil tender with Row over Gazprom Tax', *BBC Summary of World Broadcasts*, 7 July 1998. A similar linkage is also suggested by 'Russia: Shelling Out', *Economist Intelligence Unit: Crossborder Monitor*, 15 July 1998.

38. Vyacheslav Nikonov, 'Pravitel'stvo – "Gazprom" 1:1', *Izvestiya*, 7 July 1998.

39. In game theoretic terms, 'private information' refers to a situation of information asymmetry, whereby privileged information accrues only to a particular actor usually as a result of his particular position, for instance, as a company manager or as a member of government. In this respect, other actors are said to possess incomplete information.

40. On 13 July 1998, the IMF approved US$17.1 billion in new loans to Russia, of which US$4.8 billion was disbursed at the end of July.

41. These figures are from *Russian Economy in 1998*; and Freeland, *Sale of the Century*, p. 241.

42. Prior to 1997, debt servicing took up 25% of each ruble borrowed. In 1997, this figure rose to 85%, and in 1998 all new borrowings were used to repay old debts. See Thierry Malleret, Natalia Orlova and Vladimir Romanov, 'What Loaded and Triggered the Russian Crisis?', *Post-Soviet Affairs* 15, no. 2 (1999), p. 111. For other analysis of the crisis in Russia, see *Russian Economy in 1998*; and Clifford Gady and Barry Ickes, 'Russia's Virtual Economy', *Foreign Affairs* 77, no. 5 (1998): 53–67.

43. Prices are for Urals oil as cited in *Energy Prices & Taxes: 1st Quarter 1998* (Paris: International Energy Agency, 1998), p. 32; and *Energy Prices & Taxes: 3rd Quarter 1998* (Paris: International Energy Agency, 1999), p. 4.

44. This is again reflective of the tendency, mentioned earlier in the book, of blaming big business for the shortcomings in economic policy, especially of privatization. See, for instance, Nina Pusenkova, '"Rosneft"' kak zerkalo russkoy evolyutsii', *Pro et Contra* 10, no. 2–3 (2006): 91–103.

45. Candidates for the post of president of Rosneft' included former Mintopenergo heads Yuriy Shafranik and Sergey Chizov as well as former GKI heads Chubays and Kazakov.

46. From an interview with Bogdanchikov by Evgeniy Bogin, 'Sergey Bogdanchikov: "O planakh privatizatsii 'Rosnefti' nado na nekotoroe vremya zabyt'"', *Profil*, 26 October 1998.
47. See interview by Magiy Ignatovoy, 'Sergey Bogdanchikov, prezident gosudarstvennoy neftyanoy kompanii "Rosnefti": ne nravitsya mne etot shum v otrasli', *Izvestiya*, 8 September 2003.
48. For a background of the struggle to maintain control over these subsidiaries, see Nikolay Poluektov and Andrey Bagrov, 'Operatsiya "Purneftegaz"', *Kommersant'-Daily*, 7 October 1998; 'Rosneft Regains Purneftegaz but Remains Short of Funds', *Financial Times Energy Newsletters: East European Energy Report*, 1 October 1998; Gary Peach, 'The Death of Rosneft', *Moscow Times*, 1 December 1998; 'Dyrka ot skvazhiny', *Novaya gazeta*, 16 December 2002; 'Sovladel'tsy pustoty', *Novaya gazeta*, 12 May 2003; 'Neftyanaya vertikal' kompaniya "Rosneft'" "smogla vernut" gosudarstvu uvedennye aktivy', *Novaya gazeta*, 29 April 2004; and Mikhail Yenukov, 'Kusochki "Rosnefti"', *Vedomosti*, 13 August 2000.
49. Aleksey Makarkin and Sergey Savushkin, 'Rol' lichhosti v istorii neftyanoy goskompanii', *Neft' i Kapital*, no. 10, 2002.
50. Shevtsova, 'Prezident Putin oformlyaet sobstvennyy politicheskiy rezhim'.
51. Rosneft' has faced many problems in operating in Chechnya, particularly during the tenure of the region's plenipotentiary of the federal government at that time, Nikolay Koshman, who refused to cede control over the oil industry in view of his vested interests in illegal oil sales. See, for example, Milana Davydova and Yuliya Ul'yanova, 'Koshman oyl', *Segodnya*, 20 June 2000. Since 1991, rival Chechen clans have been fighting to control the republic's oil resources, and they have found support among segments of the Russian military there. For a brief review, see 'Chechen Oil Provoked War', *Moscow News*, 4 July 2001.
52. The figures are from a variety of sources, including *Energy Policies of the Russian Federation*, p. 50; and *Russia Energy Survey 2002*, p. 144. The merger would also liberalize trading in Gazprom's shares by raising the current limit on foreign ownership, thereby significantly increasing the value of the state's own stake in Gazprom.
53. Quote from an independent consultancy firm as cited by Guzel Fazullina, 'State Officials Are Going into Business', *Rodnaya gazeta*, 12 November 2004 (printed in *Johnson's Russia's List* no. 8449).
54. 'Gazprom–Rosneft Merger Runs into Personal Rivalries', (Moscow: Rye, Man & Gor Securities, 4 October 2004). For a profile of the different members of each 'faction' and their political outlook, see the literature cited in Chapter 4. Some analysts, however, have questioned the extent of actual policy differences between the two 'factions'. See Greg Walters, 'Study: Siloviki's Struggle for Assets Not Over', *Moscow Times*, 27 June 2005. The posts cited in the text were those prior to May 2008. Since then, Dmitriy Medvedev has become the President of Russia.
55. There is some evidence that Miller and Medvedev have pursued personal advantages. For example, in 2003, Gazprom lost US$2.1 billion in profits due to questionable practices and expenditure, including almost US$800 million from contracting out gas-supply operations on highly favourable terms to a company suspected of having close links to Gazprom's top management.

Under the previous management of Gazprom, led by Rem Vyakhirev, the company lost about US$3 billion annually and was similarly accused of contracting out gas-supply operations to a company with known links to Gazprom. This company Itera has since ceased to exist, but many analysts think that its place and role has simply been replaced by Eural Trans Gas under Medvedev and Miller. See Irina Reznik, 'Gazprom poteryal $2.1 mlrd', *Vedomosti*, 9 June 2004; Catherine Belton, 'Gazprom Leaking Billions, Report Says', *Moscow Times*, 8 June 2004; William F. Browder, 'Gazprom and Itera: A Case Study in Russian Corporate Misgovernance', 18 March 2002 (printed in *Johnson's Russia's List* no. 6149); and Jason Bush, 'Murky Deals at Gazprom', *Business Week*, 21 June 2004.

56. See Yuliya Bushueva and Yekaterina Derbilova, '"Rosneft'" podorozhala', *Vedomosti*, 16 December 2004. Gazprom hired Dresdner Kleinwort Wasserstein while Rosneft' turned to Morgan Stanley. The head of Dr KW's Russia office Mr Matthias Warnig was recently nominated to Gazprom's board of directors and has known Putin since the latter's KGB stint in Germany. See Guy Chazan and David Crawford, 'In From the Cold', *Wall Street Journal*, 23 February 2005.

57. The injunction was overturned in February 2005, but it came too late for Gazprom and its subsidiaries to participate in the auction for Yuganskneftegaz.

58. Quote cited by Petr Sapozhnikov, 'My prodolzhaem construktivnoe sotrudnichestvo s YUKOSom', *Kommersant'-Daily*, 7 February 2005. Nevertheless Gazprom's Miller continued to insist that the state would obtain a controlling stake in Gazprom in exchange for 100% of Rosneft' minus Yuganskneftegaz, which he opined would become a separate state-owned entity. See Nikolay Gorelov, 'Slivalis' dva tovarishcha', *Vremya novostey*, 3 March 2005.

59. See, for example, '. . . as former Finance Minister speculates on State's plans for Yukos', *Radio Free Europe/Radio Liberty Newsline*, 23 July 2004; Andrew Osborn, 'Kremlin Ally Bogdanov Emerges as Buyer of key Yukos subsidiary', *The Independent (London)*, 22 December 2004.

60. Mikhail Gorbachev, the last president of the Soviet Union, endorsed Putin's decision to launch an IPO for Rosneft' because it gave 'many more ordinary Russians the opportunity to have a direct stake in the future success of the Russian economy'. As cited by Joanna Chung and Arkady Ostrovsky, 'Rosneft IPO Fails to Attract Big Players', *Financial Times*, 15 July 2006.

61. For details of the scheme, see Ekaterina Derbilova, Irina Reznik and Andrey Panov, 'Milliardy dlya "Gazproma",' *Vedomosti*, 18 May 2005.

62. Quote by Vladimir Milov, an oil analyst and former deputy minister, in Catherine Belton, 'An IPO Built on Greed and Ambition', *Moscow Times*, 7 July 2006.

63. From an unnamed banker as cited in ibid.

64. See 'BP, Petronas, CNPC take up Stakes in Rosneft IPO', *Platts Energy Economist*, 1 August 2006. China National Petroleum Company expressed an interest to purchase up to US$3 billion of shares at the IPO or around 4.5% of Rosneft', but were only allocated US$500 million or around 0.7% of Rosneft'.

65. The American investment banker George Soros and executives of YUKOS were among those who campaigned against participation in the Rosneft' IPO. See Neil Buckley, 'Soros says Rosneft IPO Raises Concerns', *Financial*

Times, 26 April 2006; and Alexander Temerko, 'Can We Trust Rosneft's Current Owner, the Russian State?', *Wall Street Journal Europe*, 24 May 2006.

66. See 'Glava "Rosnefti" poka na meste', *Kommersant'-Daily*, 18 September 1999; Petr Sapozhnikov, 'Nachalo "Gosnefti": "Slavneft" pomenyala prezidenta', *Kommersant'-Daily*, 15 January 2000; and Nikolay Ivanov, '"Ili rabotat' . . . "' *Segodnya*, 20 October 1999.

67. Based on production figures for 2005.

68. Abramovich purchased US$300 million worth of shares, according to Ekaterina Derbilova and Mariya Rozhkova, 'Tri Kita "Rosnefti"', *Vedomosti*, 24 July 2006.

69. This sentiment was expressed by BP, which paid US$1billion for a 1.4% stake in Rosneft'. With regard to the price per share for Rosneft', a fund manager noted that 'you have to be a real idiot to buy something trading on a premium to LUKoil on a price-per-earnings basis'. From Catherine Belton, 'Half of Rosneft IPO Goes to 4 Buyers', *Moscow Times*, 17 July 2006.

70. See 'President Delays Sale of Rosneft Stake', *Moscow Times*, 6 September 2000; and Irina Reznik, 'Inostrantsy boyatsya razdela informatsii vmesto razdela produktsii', *Kommersant'-Daily*, 3 September 2001. Production-sharing projects with the involvement of Rosneft' are cited in 'Downer: Rosneft Gets Reduced PSA Role', *Nefte Compass*, 10 April 2002. However, only a handful of PSA projects are actually operational at this point in time. See also 'Putin Pushes Rosneft into the Limelight', *Petroleum Intelligence Weekly*, 27 June 2002.

71. The criticism is by former presidential economic adviser Andrey Illarionov, as cited in Catherine Belton, 'Illarionov Slams Rosneft IPO', *Moscow Times*, 5 July 2006.

72. See Aleksandr Tutushkin, Ekaterina Derbilova and Vera Surzhenko, 'Eshche posluzhit', *Vedomosti*, 24 July 2006.

73. The appointment of Sechin as chairman of Rosneft' was contrary to the hitherto practice of naming a senior state bureaucrat to that position: the former chairpersons of Rosneft' included German Gref (Minister of Economy) and Igor' Yusufov (former head of Mintopenergo). Sergey Oganesyan, as head of the Federal Energy Agency and a former vice-president of Rosneft', was widely expected to be named the new chairman in July 2004. See Dmitriy Butrin, Petr Sapozhnikov and Mariya Molina, 'Neftyanoy kardinal', *Kommersant'-Daily*, 28 July 2004. For a list of officials from the presidential administration serving on the boards of state-owned companies, see Fazullina, 'State Officials Are Going into Business'.

74. As argued by Nodari Simonia, 'Russia at the Turning Point', 17 September 2004 (printed in *Johnson's Russia's List* no.8372).

75. The fact that Bogdanchikov's son was appointed in February 2006 as the Head of Investor Relations at Rosneft' prior to the IPO, and that the son of a leading *siloviki*, Nikolay Patrushev (director of the Federal Security Service) was given a post in September 2006 as advisor to the chairman of Rosneft', appears to lend credence to the view that Rosneft' is the domain of the *siloviki* whereas Gazprom is controlled by the 'petersburgers'. See Yelena Kiseleva, Dmitriy Butrin and Mikhail Fishman, '"Rosneft"' vo vtorom pokolenii', *Kommersant'-Daily*, 13 September 2006.

76. Denis Skorobogat'ko, Dmitriy Butrin, Irina Rybal'chenko and Elena Kiseleva, 'Gosudarstvo raskololos' na neft' i gaz', *Kommersant'-Daily*, 2 February 2005.

77. From Aleksey Grivach, 'Obyknovennyy lobbizm: partiya "Rosnefti" igraet na publiku', *Vremya novostey*, 2 February 2005.
78. See 'Yuganskneftegaz ne otdadut "Gazpromu"', *Vedomosti*, 17 March 2005.
79. Mikhail Khodorkovskiy, 'Tyur'ma i mir: sobstvennost' i svoboda', *Vedomosti*, 28 December 2004.

6 Conclusion

1. The exceptions are LUKoil and Surgutneftegaz, where the Soviet-era general managers have control and ownership of these companies.
2. Timothy Frye, 'Capture or Exchange? Business Lobbying in Russia', *Europe–Asia Studies* 54, no. 7 (2002), p. 1031.
3. S. P. Peregudov, 'Krupnaya rossiyskaya korporatsiya v sisteme vlasti', *Polis*, no. 3 (2001), p. 18.
4. Gaidar, *State & Evolution*, p. 90. In contrast, post-1989 East Central Europe introduced lustration laws against the *nomenklatura*. Hence, changes in Russia may be regarded as anti-Soviet (that is, aimed at dissolving the Soviet Union itself) rather than anti-communist in nature. See Hilary Appel, 'The Ideological Determinants of Liberal Economic Reform: The Case of Privatisation,' *World Politics* 52, no. 4 (2000): 520–549.
5. Since 1999, Fridman's Al'fa group has expanded into the telecommunications and supermarket sectors, while Potanin has bought out his partner and now owns 55% of Norilsk Nickel, one of the world's largest nickel producer. For more information, see Fortescue, *Oil Barons and Metal Magnates*.
6. John W. Kingdon, *Agendas, Alternatives, and Public Policies* (New York: Longman, 1995), p. 165.
7. Churilov, the Soviet Oil Minister, claimed that reforms to create a few VIOCs were 'looming in the air' whereas a former deputy minister of Mintopenergo argued that a single, national VIOC modelled on Gazprom was a more likely outcome in the late 1980s and early 1990s. From author's interviews with Mr Lev Churilov and Mr Andrey Konoplyanik, respectively.
8. Author's interview with Mr Viktor Chernomyrdin.
9. Frank R. Baumgartner and Bryan D. Jones, 'Agenda Dynamics and Policy Subsystems', *The Journal of Politics* 53, no. 4 (1991), p. 1047.
10. Anders Aslund, 'The Russian President's Second Term Disaster', *The Weekly Standard*, 17 January 2005.
11. According to the provisions of the TNK–BP deal of 2004, the Russian owners would be free to dispose of their shares only after three years, that is, in 2007. It is widely assumed that these shares will be sold to Gazprom. For Slavneft', which is currently owned in equal parts of 49.48% by TNK and BP, such a scenario would mean that it will no longer be a fully privately owned oil company, but one owned in part by state-owned Gazprom. See Irina Reznik and Tat'yana Yegorova, 'Gazprom-BP', *Vedomosti*, 18 September 2006; and Tat'yana Yegorova and Irina Reznik, '"Gazprom" prismatrivaetsya k "Slavnefti"', *Vedomosti*, 21 November 2005.
12. Quote by the general manager of Kommersant' group as cited in 'Vasil'ev: "Al'fa-Bank" pytaetsya razorit' "Kommersant", no emu eto ne udastsya', *Lenta.ru*, 20 October 2004. Al'fa won a court order for US$11.7 million in damages from the Kommersant' group, which is owned by Berezovskiy.

13. From Robert Coalson, 'The Assault on "Kommersant"', *Radio Free Europe/ Radio Liberty Media Matters*, 25 October 2004.
14. John Kenneth Galbraith, 'Countervailing Power', *American Economic Review: Papers and Proceedings of the 66th Annual Meeting of the American Economic Association* 44, no. 2 (1954): 1–6.
15. 'Who will the Next Duma Really Serve?', *Moscow Times*, 4 December 2003.
16. Clifford Gaddy and Barry Ickes, *Russia's Virtual Economy* (Washington DC: Brookings Institution Press, 2002), p. 56. Yevgeniy Shvidler, the president of Sibneft', similarly noted that 'every large oil company possesses certain acquaintances, administrative resources and so on, which were acquired by it during the long years of working in the Russian marketplace'. As quoted by Peregudov, 'Krupnaya rossiyskaya koporatsiya v sisteme vlasti', p. 18.
17. See Clifford G. Gaddy, 'Perspectives on the Potential of Russian Oil', *Eurasian Geography and Economics* 45, no. 5 (2004), p. 350, footnote 6.
18. From Peter Rutland, Gail W. Lapidus, Barry Ickes, Carol Saivetz and George W Breslauer, 'Russia in the Year 2003', *Post-Soviet Affairs* 20, no. 1 (2004), pp. 30–31. Abramovich's decision in recent years to cash-out of his holdings in Russia may also reflect his realization that his stock of relational capital could be depreciating.
19. See Aleksandr Igorev, Sergey Topol', Petr Sapozhnikov and Irina Reznik, 'Prishli za Alekperovym', *Kommersant'-Daily*, 12 July 2000; and Viktoriya Abramenko, Milana Davydova and Georgiy Osipov, 'Sezon okhoty na krupnyy biznes', *Segodnya*, 12 July 2000. The charges were dismissed one month later.
20. In 2005, Gazprom's consultants proposed that Surgutneftegaz, along with Yuganskneftegaz and Rosneft', be acquired as Gazprom's oil subsidiaries. There were also rumours in 2006 that Rosneft' would acquire Surgutneftegaz. See Tat'yana Yegorova, Irina Reznik, Yekaterina Derbilova and Mariya Rozhkova, '"Surgut" ne slivaetsya', *Vedomosti*, 31 January 2006.
21. Likewise, the fact that Tatneft' and Bashneft' are state-owned is less convincing than the objective explanation that their higher levels of drilling are necessitated by the older oil fields in Tartarstan and Bashkortostan, compared to west Siberia. Information on investment levels by company is from Troika Dialog's *Oil Sector Report*, p. 41.
22. The point that the business model, rather than relational capital, guides decision-making is also made by Michael Bradshaw and Andrew R. Bond, 'Crisis Amid Plenty Revisited: Comments on the Problematic Potential of Russian Oil', *Eurasian Geography and Economics* 45, no. 5 (2004): 352–358. The following discussion on the business models of selected Russian oil companies draws heavily from research material by two leading investment banks in Russia, Renaissance Capital and Troika Dialog.
23. From Yuliya Bushueva and Svetlana Novolodskaya, 'Alekperov raskryl karty', *Vedomosti*, 26 July 2002.
24. The data on well productivity is for 2002 to 2004 from *Russia Oil & Gas Yearbook 2003: Piping Growth* (Moscow: Renaissance Capital, July 2003), p. 25; *Russia Oil & Gas Yearbook 2004: Counting Barrels* (Moscow: Renaissance Capital, July 2004), p. 29; and *Russia Oil & Gas Yearbook 2005: Impasse* (Moscow: Renaissance Capital, July 2005), p. 29.

25. As cited by Leslie Dienes, 'Observations on the Problematic Potential of Russian Oil and the Complexities of Siberia', *Eurasian Geography and Economics* 45, no. 5 (2004), p. 323, footnote 11.
26. Interview with Khodorkovskiy as cited by Manfred F. R. Kets de Vries, Stanislav Shekshnia, Konstantin Korotov and Elizabeth Florent-Treacy, *The New Russian Business Leaders* (Cheltenham: Edward Elgar, 2004), p. 162.
27. As noted by *Oil Sector Report*, p. 42. This mentality also lies behind the Soviet/Russian method of calculating reserves. This looks at the absolute volume of reserves as opposed to the method used by Western companies, which emphasize recoverable reserves. See *Russia Energy Survey 2002*, pp. 70–71.
28. Concrete examples include the purchase of NIKoil (by LUKoil) and Severnaya Neft' (by Rosneft'). Also, around 15% of LUKoil is held by foreign shareholders, and in 2000, LUKoil purchased the Getty Petroleum marketing chain in the USA. As for Rosneft', it has the largest number of production-sharing projects with foreign partners.
29. Hall and Taylor, 'Political Science and the Three New Institutionalisms', p. 945.
30. Between 1993 and 1997, implicit subsidies provided by Gazprom amounted to US$5.5 billion per year on average, or 1.6% of Russia's GDP. Likewise, the electricity company, Unified Energy Systems, provided implicit subsidies of US$7.4 billion per year on average or 2.3% of GDP. The data are from Bobylev and Cukrowski, 'Russia: Bank Assistance for the Energy Sector', pp. 37–38. Gazprom's integral role in this respect was also noted by Gaddy and Ickes, 'Russia's Virtual Economy'.
31. Quote from Nemtsov from 'First Deputy Premier Nemtsov Admits No-One Knows where Gazprom's Money Goes', *BBC Summary of World Broadcasts*, 15 April 1997.
32. Author's interview with Mr Nick Butler.
33. As noted by Peters, *Institutional Theory in Political Science*, p. 47.
34. The head of Rosugol' was Dr Yuriy Malyshev, who continued in his capacity until the dissolution of the organization in 1997. The other reason why it persisted till December 1997 was that it served the interests of the relevant actors, as mentioned earlier.
35. Andrei Shleifer and Sergey Vasiliev, 'Management Ownership and Russian Privatisation', in *Corporate Governance in Central Europe and Russia: Insiders and the State*, ed. Roman Frydman, Cheryl W. Gray and Andrzej Rapaczynski (Budapest: Central European Univeristy Press, 1996), p. 68.
36. Shleifer and Boyko, 'The Politics of Russian Privatisation', p. 55.
37. Putin, 'Russia at the Turn of the Millennium'.

Bibliography

Interviews

Mr Nick Butler, Group vice-president (Strategy), British Petroleum. 20 January 2005, London.

Mr Viktor Chernomyrdin, former chairman of Gazprom and ex-prime minister of the Russian Federation. 11 February 2004, Kiev.

Mr Lev Churilov, former minister of Oil of the USSR. 12 February 2004, Moscow.

Mr Anatoliy Fomin, former deputy minister of Fuel & Energy and ex-president of Slavneft'. 9 February 2004, Moscow.

Mr Sergey Generalov, former vice-president of YUKOS and ex-minister of Fuel & Energy. 30 September 2003, Moscow.

Mr Jonathan Hay, former head of Harvard Institute for International Development (Moscow office). 17 November 2004, London.

Mr Viktor Ivanenko, former acting chairman of the KGB and ex-vice-president of YUKOS. 6 February 2004, Moscow.

Dr Andrey Konoplyanik, former deputy minister of Fuel & Energy. 28 June 2004, Brussels.

Mr Vladimir Lopukhin, former minister of Fuel & Energy. 3 February 2004, Moscow.

Mr Sergey Molozhaviy, former deputy minister of GKI. 10 February 2004, Moscow.

Mr Viktor Ott, former vice-president of Rosneft' and ex-deputy minister of Fuel & Energy. 14 February 2004, Moscow.

Mr Aleksandr Putilov, former president of Rosneft'. 3 October 2003, Moscow.

Mr Dmitriy Romanov, ex-dDeputy minister of GKI and former vice-president of Slavneft'. 3 February 2004, Moscow.

Mr Yuriy Shafranik, former minister of Fuel and Energy. 13 February 2004, Moscow.

Mr Andrey Shtorkh, former vice-president of Slavneft'. 1 October 2003.

Dr Rair Simonyan, former first vice-president of Rosneft'. 1 October 2003, Moscow.

Dr Rair Simonyan, former first vice-president of Rosneft'. 23 May 2002, Oxford.

Books/Book chapters

Aage, Hans. 'Privatisation and Democratisation in Eastern Europe and the Former Soviet Union'. In *International Privatisation: Strategies and Practices*, edited by Thomas Clarke. Berlin: Walter de Gruyter, 1994.

Allison, Graham T. *Essence of Decision: Explaining the Cuban Missile Crisis*. Boston: Little Brown, 1971.

Almond, Gabriel A. and G. Bingham Powell. *Comparative Politics: A Developmental Approach*. Boston: Little Brown, 1966.

Ames, B. *Political Survival: Politicians and Public Policy in Latin America*. Berkeley: University of California Press, 1987.

Arkhipov, Sergei, Said Batkibekov, Sergei Drobyshevsky, Vladimir Mau, Sergei Sinelnikov-Murylev and Alexei Uluykaev. 'Macroeconomic Stabilisation and Fiscal Crisis'. In *The Economics of Transition*, edited by Yegor Gaidar. Cambridge, Massachusetts: MIT Press, 2002.

Aslund, Anders. *Building Capitalism: The Transformation of the Former Soviet Bloc*. Cambridge: Cambridge University Press, 2002.

——. *Russia's Economic Transformation in the 1990s*. Washington DC: Pinter, 1997.

——. *How Russia Became a Market Economy*. Washington DC: Brookings Institution Press, 1995.

Aton Capital. *Dutch Disease: Diagnosing Russia*. Moscow: Aton Capital, January 2005.

——. *No Gushers, But Good: Full-Cycle Value in Russian Oils*. Moscow: Aton Capital, April 2001.

Auty, Richard M. *Resource-Based Industrialisation: Sowing the Oil in Eight Developing Countries*. Oxford: Oxford University Press, 1990.

Balcerowicz, Leszek. *Socialism, Capitalism, Transformation*. Budapest: Central European University Press, 1995.

Ball, Alan R. and Frances Millard. *Pressure Politics in Industrial Societies: A Comparative Introduction*. Basingstoke: Palgrave Macmillan, 1986.

Barnes, Andrew. *Owning Russia: The Struggle over Factories, Farms and Power*. Ithaca: Cornell University Press, 2006.

Barton, Robert, ed. *The Almanac of Russian Petroleum 1998*. Washington DC: Energy Intelligence Group, 1998.

Bates, Robert H. 'Comment'. In *The Political Economy of Policy Reform*, edited by John Williamson. Washington DC: Institute for International Economics, 1994.

—— and Anne O. Krueger, eds, *Political and Economic Interactions in Economic Policy Reform: Evidence from Eight Countries*. Oxford: Blackwell, 1993.

——. 'Generalisations from the Country Studies'. In *Political and Economic Interactions in Economic Policy Reform: Evidence from Eight Countries*, edited by Robert H. Bates and Anne O. Krueger. Oxford: Blackwell, 1993.

Beer, Samuel. *Modern British Politics*. London: Faber and Faber, 1969.

Bentley, Arthur F. *The Process of Government: A Study of Social Pressures*. 1995 edn, New Brunswick: Transaction Publishers.

Berger, Suzanne, ed. *Organising Interests in Western Europe: Pluralism, Corporatism and the Transformation of Politics*. Cambridge: Cambridge University Press, 1981.

Berezovskii, Vladimir and Vladimir Chervyakov. 'Sectoral Production Capital: Military–Industrial Complex and Fuel and Energy Complex'. In *Post-Soviet Puzzles: Mapping the Political Economy of the Former Soviet Union*, edited by Klaus Segbers and Stephan de Spiegeleire. Baden Baden: Verlagsgesellschaft, 1995.

Bernstam, Michael and Alvin Rabushka. *Fixing Russia's Banks: A Proposal for Growth*. Stanford: Hoover Institution Press, 1998.

Blasi, Joseph, Maya Kroumova and Douglas Kruse. *Kremlin Capitalism: Privatising the Russian Economy*. Ithaca: Cornell University Press, 1997.

Boyko, Maxim, Andrei Shleifer and Robert W. Vishny. *Privatising Russia*. Cambridge, Massachusetts: MIT Press, 1995.

Brady, Rose. *Kapitalizm: Russia's Struggle to Free its Economy*. New Haven: Yale University Press, 1999.

Braun, Dietmar. 'Interests or Ideas?' In *Public Policy and Political Ideas*, edited by Dietmar Braun and Andreas Busch. Cheltenham: Edward Elgar, 1999.

—— and Andreas Busch, eds, *Public Policy and Political Ideas*. Cheltenhan: Edward Elgar, 1999.

Breslauer, George W. *Gorbachev and Yeltsin as Leaders*. Cambridge: Cambridge University Press, 2002.

British Petroleum. *BP Statistical Review of World Energy 2004*. June 2004.

Brown, Archie, ed. *The Demise of Marxism-Leninism in Russia*. Basingstoke: Palgrave Macmillan, 2004.

——, ed. *Contemporary Russian Politics: A Reader*. Oxford: Oxford University Press, 2001.

——. 'Political Leadership in Post-Communist Russia'. In *Russia in Search of Its Future*, edited by Amin Saikal and William Maley. Cambridge: Cambridge University Press, 1995.

——. 'Political Power and the Soviet State: Western and Soviet Perspectives'. In *The State in Socialist Society*, edited by Neil Harding. London: Macmillan Press, 1984.

—— and Lilia Shevtsova, eds, *Gorbachev, Yeltsin and Putin: Political Leadership in Russia's Transition*. Washington DC: Carnegie Endowment for International Peace, 2001.

Brudny, Yitzhak M. 'Continuity or Change in Russian Electoral Patterns? The December 1999–March 2001 Election Cycle'. In *Contemporary Russian Politics: A Reader*, edited by Archie Brown. Oxford: Oxford University Press, 2001.

Burns, Tom, Thomas Baumgartner and Philippe Deville. *Man, Society, Decisions: The Theory of Actor-System Dynamics for Social Scientists*. New York: Gordon and Breach, 1985.

Burton, Michael, Richard Gunther and John Higley. 'Elites and Democratic Consolidation in Latin American and Southern Europe: An Overview'. In *Elites and Democratic Consolidation in Latin American and Southern Europe*, edited by John Higley and Richard Gunther. Cambridge: Cambridge University Press, 1992.

Cassell, Mark. *How Governments Privatize: The Politics of Divestment in the United States and Germany*. Washington DC: Georgetown University Press, 2002.

Cawson, Alan. *Corporatism and Political Theory*. Oxford: Basil Blackwell, 1986.

——, ed. *Organised Interests and the State: Studies in Meso Corporatism*. London: SAGE Publications, 1985.

Chadwick, Margaret, Machiko Nissanke and David Long. *Soviet Oil Exports: Trade Adjustments, Refining Constraints and Market Behaviour*. Oxford: Oxford University Press, 1987.

Chaisty, Paul and Jeffrey Gleisner. 'The Consolidation of Russian Parliamentarism: The State Duma, 1993–8'. In *Institutions and Political Change in Russia*, edited by Neil Robinson. Basingstoke: Palgrave Macmillan, 2000.

Chaudhry, Kiren Aziz. *The Price of Wealth: Economies and Institutions in the Middle East*. Ithaca: Cornell University Press, 1997.

Chorine, Alex. Yukos: FT Energy Company Briefings. London: Financial Times Energy, 2000.

——. Lukoil: FT Energy Company Briefings. London: Financial Times Energy, 2000.

———. Surgutneftegaz: FT Energy Company Briefings. London: Financial Times Energy, 2000.

———. Slavneft: FT Energy Company Briefings. London: Financial Times Energy, 2000.

———. Rosneft: FT Energy Company Briefings. London: Financial Times Energy, 2000.

Chubays, Anatoliy. 'V poiskakh sobstvennogo puti'. In *Privatizatsiya po-rossiyskiy*, edited by Anatoliy Chubays. Moskva: Vagrius, 1999.

Clarke, Thomas, ed. *International Privatisation: Strategies and Practices*. Berlin: Walter de Gruyter, 1994.

Colander, David. 'Introduction'. In *Neoclassical Political Economy: The Analysis of Rent Seeking and DUP Activities*, edited by David Colander. Cambridge, Massachusetts: Ballinger, 1984.

Coleman, James S. *Individual Interests and Collective Action*. Cambridge: Cambridge University Press, 1986.

———. *Power and Structure of Society*. New York: W W Norton & Company, 1974.

Collier, Ruth Berins and David Collier. *Shaping the Political Arena: Critical Junctures, the Labour Movement and Regime Dynamics in Latin America*. Princeton: Princeton University Press, 1991.

Collier, David. 'The Comparative Method'. In *Political Science: The State of the Discipline II*, edited by Ada W. Finifter. Washington DC: The American Political Science Association, 1993.

Considine, Jennifer I. and William A. Kerr. *The Russian Oil Economy*. Cheltenham: Edward Elgar, 2002.

Cox, Terry. 'Democratisation and the Growth of Pressure Groups in Soviet and Post-Soviet Politics'. In *Pressure Groups*, edited by Jeremy J. Richardson. Oxford: Oxford University Press, 1993.

Crawford, Beverly and Arend Lijphart. 'Old Legacies, New Institutions: Explaining Political and Economic Trajectories in Post-Communist Regimes'. In *Liberalisation and Leninist Legacies: Comparative Perspectives on Democratic Transitions*, edited by Beverly Crawford and Arend Lijphart. Berkeley: University of California, 1997.

Dahl, Robert. *Who Governs: Democracy and Power in an American City*. 1989 edn, New Haven: Yale University Press.

———. *A Preface to Economic Democracy*. Cambridge, Massachusetts: Polity Press, 1985.

———. *Pluralist Democracy in the United States*. Chicago: Rand McNally & Company, 1967.

de Vries, Manfred F. R. Kets, Stanislav Shekshnia, Konstantin Korotov and Elizabeth Florent-Treacy. *The New Russian Business Leaders*. Cheltenham: Edward Elgar, 2004.

Dienes, Leslie, Istvan Dobozi and Marian Radetzki. *Energy and Economic Reform in the Former Soviet Union: Implications for Production, Consumption and Exports, and for the International Energy Markets*. New York: St Martin's Press, 1994.

Diskin, Iosif. *Rossiya: Transformatsiya i elity*. Moscow: Eltra, 1995.

Dowding, Keith M. *Rational Choice and Political Power*. Aldershot: Edward Elgar, 1991.

Downs, Anthony. *Inside Bureaucracy*. Prospect Heights, Illinois: Waveland Press, 1994.

Dunleavy, Patrick. 'The Bureau-Shaping Model'. In *The Policy Process: A Reader*, edited by Michael Hill. New York: Prentice Hall, 1997.

——. *Democracy, Bureaucracy and Public Choice: Economic Explanations in Political Science*. Hemel Hempstead: Harvester-Wheatsheaf, 1991.

—— and Brendan O'Leary. *Theories of the State: The Politics of Liberal Democracy*. Basingstoke: Palgrave Macmillan, 1987.

Ebel, Robert E. *Energy Choices in Russia*. Washington DC: Centre for Strategic and International Studies, 1994.

Ellman, Michael, ed. *Russia's Oil and Natural Gas: Bonanza or Curse?* London: Anthem Press, 2006.

Elster, John. *The Multiple Self*. Cambridge: Cambridge University Press, 1986.

Energy Information Administration. *International Energy Annual 2001*. EIA, US Department of Energy, 2001.

——. *Non-OPEC Fact Sheet*. EIA, US Department of Energy, June 2002.

——. *World Production of Crude Oil, Natural Gas Plant Liquids, Other Liquids and Refinery Processing: 1980–2001*. EIA, US Department of Energy, 2002.

——. *World Coal Production: 1980–2001*. EIA, US Department of Energy, 2002.

——. *World Dry Natural Gas Production: 1980–2001*. EIA, US Department of Energy, 2002.

Epstein, Edwin M. *The Corporation in American Politics*. Englewood Cliffs, New Jersey: Prentice-Hall Inc, 1969.

European Bank for Reconstruction and Development. *Transition Report 2001*. London: EBRD, 2001.

——. *Transition Report 1999*. London: EBRD, 1999.

Evans, Peter B. 'The State as Problem and Solution: Predation, Embedded Autonomy and Structural Change'. In *The Politics of Economic Adjustment: International Constraints, Distributive Conflicts and the State*, edited by Stephan Haggard and Robert R. Kaufman. Princeton: Princeton University Press, 1992.

——, Dietrich Rueschmeyer and Theda Skocpol, eds, *Bringing the State Back In*. Cambridge: Cambridge University Press, 1985.

Finifter, Ada W., ed. *Political Science: The State of the Discipline II*. Washington DC: The American Political Science Association, 1993.

Fish, M. Steven. 'Conclusion: Democracy and Russian Politics'. In *Russian Politics: Challenges of Democratisation*, edited by Zoltan Barany and Robert G. Moser. Cambridge: Cambridge University Press, 2001.

Fortescue, Stephen. *Russia's Oil Barons and Metal Magnates: Oligarchs and the State in Transition*. Basingstoke: Palgrave Macmillan, 2007.

Freedom House. *Freedom in the World*. Washington DC: Freedom House (various years).

Freeland, Chrystia. *Sale of the Century: The Inside Story of the Second Russian Revolution*. London: Little, Brown and Company, 2000.

Friedman, Jeffrey, ed. *The Rational Choice Controversy: Economic Models of Politics Reconsidered*. New Haven: Yale University Press, 1995.

Frydman, Roman, Andrzej Rapaczynski and John S. Earle. *The Privatisation Process in Russia, Ukraine and the Baltic States*. Budapest: Central European University Press, 1993.

Frye, Timothy. *Brokers and Bureaucrats: Building Market Institutions in Russia*. Ann Arbor: University of Michigan Press, 2000.

Gaddy, Clifford and Barry Ickes. *Russia's Virtual Economy.* Washington DC: Brookings Institution Press, 2002.

Gaidar, Yegor. *Collapse of an Empire: Lessons for Modern Russia.* Translated by Antonina W. Bouis. Washington DC: Brookings Institution Press, 2007.

——. *State & Evolution: Russia's Search for a Free Market.* Seattle: University of Washington Press, 2003.

——, ed. *The Economics of Transition.* Cambridge, Massachusetts: MIT Press, 2002.

——. *Days of Victory and Defeat.* Translated by Jane Ann Miller. Seattle: University of Washington Press, 1999.

Garrett, Geoffrey and Barry R. Weingast. 'Ideas, Interests and Institutions: Constructing the European Community's Internal Market'. In *Ideas and Foreign Policy: Beliefs, Institutions and Political Change,* edited by Judith Goldstein and Robert O. Keohane. Ithaca: Cornell University Press, 1993.

Geddes, Barbara. 'How the Cases You Choose Affect the Answers You Get: Selection Bias in Comparative Politics'. In *Political Analysis: An Annual Publication of the Methodology Section of the American Political Science Association,* edited by James A. Stimson. Ann Arbor: University of Michigan Press, 1990.

Gelb, Alan. *Windfall Gains: Blessing or Curse?* Oxford: Oxford University Press, 1988.

Gevorkyan, Nataliya, Natalya Timakova and Andrei Kolesnikov. *First Person: An Astonishingly Frank Portrait by Russia's President Vladimir Putin.* Translated by Catherine A. Fitzpatrick. New York: Public Affairs, 2000.

Ghadar, Fariborz. 'Oil: The Power of an Industry'. In *The Promise of Privatisation: A Challenge for US Policy,* edited by Raymond Vernon. Washington DC: Council of Foreign Relations, 1988.

Goldman, Marshall I. *The Piratisation of Russia: Russian Reform Goes Awry.* London: Routledge, 2003.

——. *Lost Opportunity: What Has Made Economic Reform in Russia so Difficult.* 2nd edn, New York: Norton, 1996.

——. *The Enigma of Soviet Petroleum: Half-Full or Half-Empty?* London: George Allen & Unwin, 1980.

Goldstein, Judith and Robert O. Keohane, eds, *Ideas and Foreign Policy: Beliefs, Institutions and Political Change.* Ithaca: Cornell University Press, 1993.

Gordon, Claire. 'Institutions, Economic Interests and the Stalling of Economic Reform'. In *Institutions and Political Change in Russia,* edited by Neil Robinson. Basingstoke: Palgrave Macmillan, 2000.

Gouliev, Rasul. *Oil and Politics: New Relationships among the Oil Producing States – Azerbaijan, Russia, Kazakhstan, and the West.* New York: Liberty Publishing House, 1997.

Grant, Jordan A. and Jeremy J. Richardson. *Government and Pressure Groups in Britain.* Oxford: Clarendon Press, 1987.

Grant, Wyn. *Business and Politics in Britain.* 2nd edn, Basingstoke: Palgrave Macmillan, 1993.

——. 'The Role and Power of Pressure Groups'. In *British Politics in Perspective,* edited by R. L. Borthwick and J. E. Spence. Leicester: Leicester University Press, 1984.

Green, Donald and Ian Shapiro. *Pathologies of Rational Choice Theory: A Critique of Applications in Political Science.* New Haven: Yale University Press, 1994.

Griffiths, Franklyn. 'A Tendency Analysis of Soviet Policy-Making'. In *Interest Groups in Soviet Politics*, edited by H. Gordon Skilling and Franklyn Griffiths. Princeton: Princeton University Press, 1971.

Grindle, Merilee S. and John W. Thomas. *Public Choices and Policy Change*. Baltimore: Johns Hopkins University Press, 1991.

——. 'Policymakers, Policy Choices and Policy Outcomes: Political Economy of Reform in Developing Countries'. In *Reforming Economic Systems in Developing Countries*, edited by Dwight H. Perkins and Michael Roemer. Cambridge, Massachusetts: Harvard University Press, 1991.

Grossman, Gene M. and Elhanam Helpman. *Special Interest Politics*. Cambridge, Massachusetts: MIT Press, 2001.

Gustafson, Thane. *Capitalism Russian-Style*. Cambridge: Cambridge University Press, 1999.

——. *Crisis Amid Plenty: The Politics of Soviet Energy under Brezhnev and Gorbachev*. Princeton, New Jersey: Princeton University Press, 1989.

Haggard, Stephan, Sylvia Maxfield and Ben Ross Schneider. 'Theories of Business and Business–State Relations'. In *Business and the State in Developing Countries*, edited by Sylvia Maxfield and Ben Ross Schneider. Ithaca: Cornell University Press, 1997.

Haggard, Stephan and Robert R. Kaufman, eds, *The Politics of Economic Adjustment: International Constraints, Distributive Conflicts and the State*. Princeton: Princeton University Press, 1992.

——. 'Introduction: Institutions and Economic Adjustment'. In *The Politics of Economic Adjustment: International Constraints, Distributive Conflicts and the State*. Princeton: Princeton University Press, 1992.

Hall, Peter A., ed. *The Political Power of Economic Ideas: Keynesianism across Nations*. Princeton: Princeton University Press, 1989.

——. 'The Role of Interests, Institutions and Ideas in the Comparative Political Economy of the Industrialised Nations'. In *Comparative Politics: Rationality, Culture and Structure*, edited by Mark Irving Lichbach and Alan S. Zuckerman. Cambridge: Cambridge University Press, 1997.

—— and David Soskice, eds *Varieties of Capitalism: The Institutional Foundations of Comparative Advantage*. Oxford: Oxford University Press, 2001.

Hart, Jeffrey A. *Rival Capitalists: International Competitiveness in the United States, Japan and Western Europe*. Ithaca: Cornell University Press, 1992.

Hayward, Jack, Brian Barry and Archie Brown, eds *The British Study of Politics in the Twentieth Century*. Oxford: Oxford University Press, 1999.

Hedlund, Stefan. *Russian's 'Market' Economy: A Bad Case of Predatory Capitalism*. London: University College of London Press, 1999.

Heritage Foundation. *Index of Economic Freedom*. Washington DC: Heritage Foundation & Wall Street Journal (various years).

Hill, Michael, ed. *The Policy Process: A Reader*. New York: Prentice-Hall, 1997.

Hodge, Graeme A. *Privatisation: An International Review of Performance*. Boulder: Westview, 2000.

Hoffman, David E. *The Oligarchs: Wealth and Power in the New Russia*. Oxford: Public Affairs, 2002.

Hood, Christopher. *Explaining Economic Policy Reversals*. Buckingham: Open University Press, 1994.

Hoopes, Stephanie M. *Oil Privatisation, Public Choice and International Forces*. Basingstoke: Palgrave Macmillan, 1997.

Hough, Jerry. *The Logic of Economic Reform in Russia*. Washington DC: Brookings Institution Press, 2001.
——. *The Soviet Union and Social Science Theory*. Cambridge, Massachusetts: Harvard University Press, 1977.
——, Evelyn Davidheiser and Susan Goodrich Lehmann. *The 1996 Presidential Election*. Washington DC: Brookings Institution Press, 1996.
—— and Merle Fainsod. *How the Soviet Union is Governed*. Cambridge, Massachusetts: Harvard University Press, 1979.
Howlett, Michael and M. Ramesh. *Studying Public Policy: Policy Cycles and Policy Subsystems*. Oxford: Oxford University Press, 2003.
Huskey, Eugene. *Presidential Power in Russia*. New York: M E Sharpe, 1999.
Ikenberry, G. John. 'The International Spread of Privatisation Policies: Inducements, Learning and "Policy Bandwagoning"'. In *The Political Economy of Public Sector Reform and Privatisation*, edited by Ezra N. Suleiman and John Waterbury. Boulder: Westview, 1990.
Institute for Economy in Transition. *Russian Economy: Trends and Perspectives*. Moscow: IET, January 2003.
——. *Russian Economy in 1998: Trends and Perspectives*. Moscow: IET, 1999.
——. *Russian Economy in 2002: Trends and Outlooks*. Moscow: IET, 2003.
International Energy Agency. *Key World Energy Statistics 2003*. Paris: IEA, 2003.
——. *Russia Energy Survey 2002*. Paris: IEA, 2002.
——. *Energy Prices & Taxes: 3rd Quarter 1998*. Paris: IEA, 1999.
——. *Energy Prices & Taxes: 1st Quarter 1998*. Paris: IEA, 1998.
——. *Energy Policies of the Russian Federation*. Paris: IEA, 1995.
International Monetary Fund. *Russian Federation: Staff Report for the 2004 Article IV Consultation*. Washington DC: IMF, August 2004.
——. *A Study of the Soviet Economy: Volume 1*. Washington DC: IMF, 1991.
——. *A Study of the Soviet Economy: Volume 2*. Washington DC: IMF, 1991.
——. *A Study of the Soviet Economy: Volume 3*. Washington DC: IMF, 1991.
Jack, Andrew. *Inside Putin's Russia*. London: Granta, 2004.
Jackson, Peter M. and Catherine Price. 'Privatisation and Regulation: A Review of the Issues'. In *Privatisation and Regulation: A Review of the Issues*, edited by Peter M. Jackson and Catherine Price. London: Longman, 1994.
Jensen, Donald D. 'How Russia is Ruled'. In *Business and the State in Contemporary Russia*, edited by Peter Rutland. Boulder: Westview Press, 2001.
Johnson, Juliet. *A Fistful of Rubles: The Rise and Fall of the Russian Banking System*. Ithaca: Cornell University Press, 2000.
Jowitt, Ken. *New World Disorder: The Leninist Extinction*. Berkeley: University of California Press, 1992.
Kahler, Miles. 'External Influence, Conditionality and the Politics of Adjustment'. In *The Politics of Economic Adjustment: International Constraints, Distributive Conflicts and the State*, edited by Stephan Haggard and Robert R. Kaufman. Princeton: Princeton University Press, 1992.
Katznelson, Ira, Mark Kesselman and Allan Draper. *The Politics of Power: A Critical Introduction to American Government*. 4th edn, Canada: Thomson Learning, 2002.
Keck, Margaret E. and Kathryn Sikkink. *Activists beyond Borders: Advocacy Networks in International Politics*. Ithaca: Cornell University Press, 1998.
Kellison, Bruce. 'Tiumen, Decentralisation and Centre–Periphery Tension'. In *The Political Economy of Russian Oil*, edited by David Lane. Lanham: Rowman and Littlefield, 1999.

Kelley, Stanley Jr. 'The Promise and Limitations of Rational Choice Theory'. In *The Rational Choice Controversy: Economic Models of Politics Reconsidered*, edited by Jeffrey Friedman. New Haven: Yale University Press, 1995.

Kikeri, Sunita K., John Nellis and Mary S. Shirley. *Privatisation: The Lessons of Experience*. Washington DC: World Bank, 1992.

Kingdon, John W. *Agendas, Alternatives, and Public Policies*. New York: Longman, 1995.

Kiser, Larry L. and Elinor Ostrom. 'The Three Worlds of Actions: A Metatheoretical Synthesis of Institutional Approaches'. In *Strategies of Political Inquiry*, edited by Elinor Ostrom. Beverley Hills, California: SAGE Publications, 1982.

Klebnikov, Paul. *Godfather of the Kremlin: Boris Berezovsky and the Looting of Russia*. New York: Harcourt, 2000.

Kokh, Alfred. *The Selling of the Soviet Empire: Politics & Economics of Russia's Privatisation – Revelations of the Principal Insider*. New York: Liberty Publishing House, 1998.

Kolesnikov, Andrey. *Neizvestnyy Chubays: stranitsy iz biografii*. Moscow: Zakharov, 2003.

Konoplyanik, Andrey. *Rossiya na formiruyushchemsya evroaziatskom energeticheskom prostranstve: problemy konkurentosposobnosti*. Moscow: Nestor, 2004.

Kryukov, Valery and Arild Moe. *The Changing Role of Banks in the Russian Oil Sector*. London: Royal Institute of International Affairs, 1998.

——. *The New Russian Corporatism? A Case Study of Gazprom*. London: Royal Institute of International Affairs, 1996.

——. *Gazprom: Internal Structure, Management Principles and Financial Flows*. London: Royal Institute of International Affairs, 1996.

Laffont, Jean-Jacques and David Martimort. *The Theory of Incentives: The Principal–Agent Model*. Princeton: Princeton University Press, 2002.

Lane, David. 'The Political Economy of Russian Oil'. In *Business and the State in Contemporary Russia*, edited by Peter Rutland. Boulder: Westview Press, 2001.

——, ed. *The Political Economy of Russian Oil*. Lanham: Rowman and Littlefield, 1999.

——. 'The Russian Oil Elite: Background and Outlook'. In *The Political Economy of Russian Oil*, edited by David Lane. Lanham: Rowman and Littlefield, 1999.

—— and Iskander Seifumulukov. 'Structure and Ownership'. In *The Political Economy of Russian Oil*, edited by David Lane. Lantham: Rowman and Littlefield, 1999.

LaPalombara, Joseph. *Interest Groups in Italian Politics*. Princeton: Princeton University Press, 1964.

Layard, Richard and John Parker. *The Coming Russian Boom: A Guide to New Markets and Politics*. New York: The Free Press, 1996.

Ledeneva, Alena V. *Russia's Economy of Favours: Blat, Networking and Informal Exchange*. Cambridge: Cambridge University Press, 1998.

Light, Margot. 'Foreign Policy Thinking'. In *Internal Factors in Russian Foreign Policy*, edited by Neil Malcolm, Alex Pravda, Roy Allison and Margot Light. Oxford: Clarendon Press, 1996.

Lindblom, Charles E. *Politics and Markets: The World's Political–Economic Systems*. New York: Basic Books, 1977.

Linz, Juan and Alfred Stepan. *Problems of Democratic Transition and Consolidation: Southern Europe, South America and Post-Communist Europe*. Baltimore: The Johns Hopkins University Press, 1996.

Lukin, Alexander. *The Political Culture of the Russian 'Democrats'*, Oxford: Oxford University Press, 2000.

Lynch, Dov. *Russian Peacekeeping Strategies in the CIS: The Cases of Moldova, Georgia and Tajikistan*. Basingstoke: Palgrave Macmillan, 2000.

Lynn, Terry Karl. *The Paradox of Plenty: Oil Booms and Petro-States*. Berkeley: University of California Press, 1997.

Malcolm, Neil, Alex Pravda, Roy Allison and Margot Light, eds. *Internal Factors in Russian Foreign Policy*. Oxford: Clarendon Press, 1996.

Margolis, Howard. 'Dual Utilities and Rational Choice'. In *Beyond Self-Interest*, edited by Jane J. Mansbridge. Chicago: University of Chicago Press, 1990.

Marsh, David and Gerry Stoker, eds *Theory and Methods in Political Science*. Basingstoke: Palgrave Macmillan, 2002.

Mau, Vladimir. *The Political History of Economic Reform in Russia, 1985–1994*. London: The Centre for Research into Communist Economies, 1996.

—— and Irina Starodubrovskaya. *The Challenge of Revolution: Contemporary Russia in Historical Perspective*. Oxford: Oxford University Press, 2001.

Maxfield, Sylvia and Ben Ross Schneider. 'Business and the State and Economic Performance in Developing Countries'. In *Business and the State in Developing Countries*, edited by Sylvia Maxfield and Ben Ross Schneider. Ithaca: Cornell University Press, 1997.

May, Iris Geva and Aaron Wildavsky. *The Policy Cycle*. Beverly Hills: Sage, 1978.

McDaniel, Tim. *The Agony of the Russian Idea*. Princeton, New Jersey: Princeton University Press, 1996.

McFaul, Michael. *Russia between Elections: What do the 1995 Results Really Mean?* Washington DC: Carnegie Endowment for International Peace, 1996.

——. *Russia's Unfinished Revolution: Political Change from Gorbachev to Putin*. Ithaca: Cornell University Press, 2001.

——, Nikolai Petrov and Andrei Ryabov. 'Introduction'. In *Between Dictatorship and Democracy: Russian Post-Communist Political Reform*, edited by Michael McFaul, Nikolai Petrov and Andrei Ryabov. Washington DC: Carnegie Endowment for International Peace, 2004.

McGowan, Francis. 'The International Spread of Privatisation'. In *International Privatisation: Strategies and Practices*, edited by Thomas Clarke. Berlin: Walter de Gruyter, 1994.

Medvedev, Roy. *Post-Soviet Russia: A Journey through the Yeltsin Era*. Translated by George Shriver. New York: Columbia University Press, 2000.

Migdal, Joel S. *Strong Societies and Weak States: State Society Relations and State Capabilities in the Third World*. Princeton: Princeton University Press, 1988.

——, Atul Kohli and Vivienne Shue, eds. *State Power and Social Forces: Domination and Transformation in the Third World*. Cambridge: Cambridge University Press, 1994.

Miles, Marc A., Edwin J. Feulner and Mary Anastasia O'Grady, eds *Index of Economic Freedom*. Washington DC: Heritage Foundation/Wall Street Journal (various years).

Miliband, Ralph. *The State in Capitalist Society*. London: Weidenfeld & Nicholson, 1969.

Moser, Nat and Peter Oppenheimer. 'The Oil Industry: Structural Transformation and Corporate Governance'. In *Russia's Post-Communist Economy*, edited by Brigitte Granville and Peter Oppenheimer. Oxford: Oxford University Press, 2001.

Mukhin, Aleksey. *Biznes-elita i gosudarstvennaya vlast':Kto vladeet Rossiey na rubezhe vekov?* Moscow: Centr Politicheskoy Informatsii, 2001.

Neary, Peter J. and Sweder van Wijnbergen, eds *Natural Resources and the Macroeconomy.* Oxford: Blackwells, 1986.

Nelson, Joan M., ed. *Economic Crisis and Policy Choice: The Politics of Adjustment in the Third World.* Princeton: Princeton University Press, 1990.

——. 'Conclusion'. In *Economic Crisis and Policy Choice: The Politics of Adjustment in the Third World,* edited by Joan M. Nelson. Princeton: Princeton University Press, 1990.

Nelson, Lynn and Irina Kuzes. *Radical Reform in Yeltsin's Russia: Political, Economic and Social Dimensions.* New York: M E Sharpe, 1995.

——. *Property to the People: The Struggle for Radical Economic Reform in Russia.* New York: M E Sharpe, 1994.

Niskanen, William A. *Bureaucracy and Representative Government.* Chicago: Aldine, 1971.

North, Douglass C. *Structure and Change in Economic History.* New York: Norton, 1981.

Nove, Alex. *Efficiency Criteria for Nationalised Industries.* London: George Allen & Unwin, 1973.

Olson, Mancur. *Power and Prosperity: Outgrowing Communist and Capitalist Dictatorships.* New York: Basic Books, 2000.

——. *The Rise and Decline of Nations: Economic Growth, Stagflation and Social Rigidities.* New Haven: Yale University Press, 1982.

——. *The Logic of Collective Action: Public Goods and the Theory of Groups.* Cambridge, Massachusetts: Harvard University Press, 1971.

Ordeshook, Peter. *Game Theory and Political Theory.* Cambridge: Cambridge University Press, 1986.

Organisation for Economic Co-operation and Development. *OECD Economic Surveys 2004: Russian Federation.* Paris: OECD, 2004.

——. *The Investment Environment in the Russian Federation: Laws, Policies and Institutions.* Paris: OECD, 2001.

Pappe, Yakov. *Oligarkhi: Ekonomicheskaya Khronika 1992–2000.* Moscow: Gosudarstvennyy universitet vysshaya shkola ekonomiki, 2000.

Peregudov, Sergei. 'The Oligarchical Model of Russian Corporatism'. In *Contemporary Russian Politics: A Reader,* edited by Archie Brown. Oxford: Oxford University Press, 2001.

——. 'Large Corporations as National and Global Players: The Case of Lukoil'. In *Explaining Post-Soviet Patchworks: Actors and Sectors in Russia between Accommodation and Resistance to Globalisation (volume 1),* edited by Klaus Segbers. Aldershot: Ashgate, 2001.

——, N. Yu Lapina and I. S. Semenenko. *Gruppy interesov i Rossiyskoe gosudarstvo.* Moscow: Editorial URSS, 1999.

Peregudov, Sergei. *Business Interest Groups in the Political Systems of USSR and Russia.* Warwick: European Public Policy Institute, 1994.

Peters, B. Guy. *Institutional Theory in Political Science: The 'New Institutionalism'.* London: Pinter, 1999.

Peuch, Jean-Christophe. 'Russian Interference in the Caspian Sea Region: Diplomacy Adrift'. In *The Political Economy of Russian Oil,* edited by David Lane. Lanham: Rowman and Littlefield, 1999.

Pistor, Katharina. 'Company Law and Corporate Governance in Russia'. In *The Rule of Law and Economic Reform in Russia*, edited by Katharina Pistor and Jeffrey D. Sachs. Boulder: Westview Press, 1997.

Pleines, Heiko. 'Corruption and Crime in the Russian Oil Industry'. In *The Political Economy of Russian Oil*, edited by David Lane. Lanham: Rowman and Littlefield, 1999.

Politkovskaya, Anna. *Putin's Russia*. Translated by Arch Tait. London: Harvill Press, 2004.

Pravda, Alex. 'Foreign Policy'. In *Developments in Russian Politics 5*, edited by Stephen White, Alex Pravda and Zvi Gitelman. Basingstoke: Palgrave Macmillan, 2001.

Primakov, Yevgeny. *Russian Crossroads: Towards the New Millennium*. Translated by Felix Rosenthal. New Haven: Yale University Press, 2004.

Przeworski, Adam. 'Methods of Cross-National Research, 1970–1983: An Overview'. In *Comparative Policy Research: Learning from Experience*, edited by Meinolf Dierkes, Hans N. Weiler and Ariane Berthoin Antal. Brookfield: Gower, 1987.

——— and Henry Tenue. *The Logic of Comparative Social Inquiry*. New York: John Wiley, 1970.

Radygin, Alexander. *Privatisation in Russia: Hard Choice, First Results, New Targets*. London: The Centre for Research into Communist Economies, 1995.

Reddaway, Peter and Dmitri Glinski. *The Tragedy of Russia's Reforms: Market Bolshevism against Democracy*. Washington DC: United States Institute of Peace, 2001.

Renaissance Capital. *Company Handbook: January–June 2004*. Moscow: Renaissance Capital, 2004.

———. *Russian Oil & Gas Yearbook 2004: Counting Barrels*. Moscow: Renaissance Capital, July 2004.

———. *A Taxing Time for Oil*. Moscow: Renaissance Capital, January 2004.

———. *Russian Oil & Gas Yearbook 2003: Piping Growth*. Moscow: Renaissance Capital, July 2003.

———. *The Russian Budget & Its Sensitivity to Oil Prices*. Moscow: Renaissance Capital, April 2003.

———. *Slavneft and Megionneftegaz: Privatisation Implications*. Moscow: Renaissance Capital, October 2002.

Richardson, Jeremy J. 'Introduction: Pressure Groups and the Government'. In *Pressure Groups*, edited by Jeremy J. Richardson. Oxford: Oxford University Press, 1993.

Robinson, Neil, ed. *Institutions and Political Change in Russia*. Basingstoke: Palgrave Macmillan, 2000.

Roeder, Philip G. 'Transitions from Communism: State-Centred Approaches'. In *Can Democracy Take Root in Post-Soviet Russia?*, edited by Harry Eckstein, Frederic J. Jr Fleron, Erik P. Hoffman and William M. Reisinger. Lanham: Rowman and Littlefield, 1998.

Rose-Ackerman, Susan. *Corruption and Government: Causes, Consequences and Reform*. Cambridge: Cambridge University Press, 1999.

Rueschmeyer, Dietrich and Peter B. Evans. 'The State and Economic Transformation: Towards an Analysis of the Conditions Underlying Effective Intervention'. In *Bringing the State Back In*, edited by Peter B. Evans, Dietrich

Rueschmeyer and Theda Skocpol. Cambridge: Cambridge University Press, 1985.

Rutland, Peter. *Business Elites and Russian Economic Policy*. London: Royal Institute of International Affairs, 1992.

———. 'Introduction: Business and the State in Russia'. In *Business and the State in Contemporary Russia*, edited by Peter Rutland. Boulder: Westview Press, 2001.

———. 'Oil, Politics and Foreign Policy'. In *The Political Economy of Russian oil*, edited by David Lane. Lanham: Rowman and Littlefield, 1999.

———. *Lost Opportunities: Energy and Politics in Russia*. Seattle: National Bureau of Asian Research, 1997.

Sabatier, Paul A., ed. *Theories of the Policy Process*. Boulder, Colorado: Westview Press, 1998.

——— and Hank C. Jenkins-Smith. 'The Advocacy Coalition Framework: An Assessment'. In *Theories of the Policy Process*. Boulder, Colorado: Westview Press, 1998.

Sachs, Jeffrey D. 'Why Russia Has Failed to Stabilise'. In *Russia's Economic Transformation in the 1990s*, edited by Anders Aslund. Washington DC: Pinter, 1997.

Salomon Brothers. *Prospects for Progress: Industry Background and Status*. London: Salomon Brothers, March 1996.

———. *Matriarchs of the New Millennium: Vertically Integrated Company Profiles*. Solomon Brothers, March 1996.

Satter, David. *Darkness at Dawn: The Rise of the Russian Criminal State*. New Haven: Yale University Press, 2003.

Scharpf, Fritz W. *Games Real Actors Play: Actor-Centred Institutionalism in Policy Research*. Boulder: Westview Press, 1997.

Schattschneider, Elmer E. *The Semi-Sovereign People*. New York: Holt, Rinehart and Winston, 1970.

Schlager, Edella. 'A Comparison of Frameworks, Theories and Models of the Policy Process'. In *Theories of the Policy Process*, edited by Paul A. Sabatier. Boulder, Colorado: Westview Press, 1999.

Schmitter, Philippe C. 'Interest Intermediation and Regime Governability in Contemporary Western Europe and North America'. In *Organising Interests in Western Europe: Pluralism, Corporatism and the Transformation of Politics*, edited by Suzanne Berger. Cambridge: Cambridge University Press, 1981.

——— and Gerhard Lehmbruch, eds *Patterns of Corporatist Policy-Making*. London: SAGE Publications, 1982.

———, eds. *Trends towards Corporatist Intermediation*. London: SAGE Publications, 1979.

Segbers, Klaus and Stephan de Spiegeleire, eds, *Post-Soviet Puzzles: Mapping the Political Economy of the Former Soviet Union*. Baden Baden: Verlagsgesellschaft, 1995.

Shafer, D. Michael. *Winners and Losers: How Sectors Shape the Developmental Prospects of States*. Ithaca: Cornell University Press, 1994.

Shafranik, Yuriy and Valeriy Kryukov. *Zapadno-Sibirskiy Fenomen*. Moscow: Neftegazovaya vertikal', 2000.

Shepsle, Kenneth A. 'Comment on Derthick and Quirk'. In *Regulatory Policy and the Social Sciences*, edited by Roger G. Noll. Berkeley: University of California Press, 1985.

—— and Mark S. Bonchek. *Analysing Politics: Rationality, Behaviour and Institutions*. New York: W W Norton & Company, 1997.

Shevtsova, Lilia. *Putin's Russia*. Washington DC: Carnegie Endowment for International Peace, 2003.

——. 'From Yeltsin to Putin: The Evolution of Presidential Power'. In *Gorbachev, Yeltsin and Putin: Political Leadership in Russia's Transition*, edited by Archie Brown and Lilia Shevtsova. Washington DC: Carnegie Endowment for International Peace, 2001.

——. *Yeltsin's Russia: Myths and Reality*. Washington DC: Carnegie Endowment for International Peace, 1999.

Shleifer, Andrei and Daniel Treisman. *Without a Map: Political Tactics and Economic Reform in Russia*. Cambridge, Massachusetts: MIT Press, 2000.

—— and Maxim Boyko. 'The Politics of Russian Privatisation'. In *Post-Communist Reform: Pain and Progress*, edited by Oliver Blanchard, Maxim Boyko, Marek Dabrowski, Richard Layard and Andrei Shleifer. Cambridge, Massachusetts: MIT Press, 1993.

—— and Sergey Vasiliev. 'Management Ownership and Russian Privatisation'. In *Corporate Governance in Central Europe and Russia: Insiders and the State*, edited by Roman Frydman, Cheryl W. Gray and Andrzej Rapaczynski. Budapest: Central European University Press, 1996.

Sikkink, Kathryn. *Ideas and Institutions: Developmentalism in Brazil and Argentina*. Ithaca: Cornell University Press, 1991.

Simonia, Nodari. 'Economic Interests and Political Power in Post-Soviet Russia'. In *Contemporary Russian Politics: A Reader*, edited by Archie Brown. Oxford: Oxford University Press, 2001.

Skocpol, Theda. *States and Social Revolutions: A Comparative Analysis of France, Russia and China*. Cambridge: Cambridge University Press, 1979.

Solnick, Steven Lee. *Stealing the State: Control and Collapse in Soviet Institutions*. Cambridge, Massachusetts: MIT Press, 1998.

Stallings, Barbara. 'International Influence on Economic Policy: Debt, Stabilisation and Structural Reform'. In *The Politics of Economic Adjustment: International Constraints, Distributive Conflicts and the State*, edited by Stephan Haggard and Robert R. Kaufman. Princeton: Princeton University Press, 1992.

Starr, Paul. 'The New Life of the Liberal State: Privatisation and the Restructuring of State–Society Relations'. In *The Political Economy of Public Sector Reform and Privatisation*, edited by Ezra N. Suleiman and John Waterbury. Boulder: Westview, 1990.

——. 'The Meaning of Privatisation'. In *The Retreat of the State: Deregulation and Privatisation in the UK and the US*, edited by Sheila B. Kamerman, Alfred J. Kahn and Dennis Swann. Ann Arbor, Michigan: University of Michigan Press, 1988.

Stinchcombe, Arthur. *Theoretical Methods in Social History*. New York: Academic Press, 1978.

Suleiman, Ezra N. and John Waterbury, eds, *The Political Economy of Public Sector Reform and Privatisation*. Boulder: Westview Press, 1990.

——. 'Introduction: Analysing Privatisation in Industrial and Developing Countries'. In *The Political Economy of Public Sector Reform and Privatisation*. Boulder: Westview Press, 1990.

Tchurilov, Lev. *Lifeblood of Empire: A Personal History of the Rise and Fall of the Soviet Oil Industry*. New York: PIW Publications, 1996.

Thomas, Clive S. 'Understanding and Comparing Interest Groups in Western Democracies'. In *First World Interest Groups: A Comparative Perspective*, edited by Clive S. Thomas. Westport, Connecticut: Greenwood Press, 2000.

Timofeyev, Igor V. 'The Development of Russian Liberal Thought since 1985'. In *The Demise of Marxism–Leninism in Russia*, edited by Archie Brown. Basingstoke: Palgrave Macmillan, 2004.

Transparency International. *Corruption Perception Index*. Berlin: Transparency International (various years).

Troika Dialog. *Oil Sector Report*. Moscow: Troika Dialog, May 2001.

——. *Oil Sector Report*. Moscow: Troika Dialog, March 2000.

Troxel, Tiffany A. *Parliamentary Power in Russia, 1994–2001: President versus Parliament*. Basingstoke: Palgrave Macmillan, 2003.

Truman, David. *The Governmental Process: Political Interests and Public Opinion*. New York: Alfred A Knopf, 1951.

Tullock, Gordon. *The Vote Motive*. London: Institute for Economic Affairs, 1975.

Ulyukaev, Aleksei. *Reforming the Russian Economy*. London: Centre for Research into Post-Communist Economies, 1996.

United Nations. *World Investment Report 2003*. New York: United Nations, 2003.

Varese, Frederico. *The Russian Mafia: Private Protection in a New Market Economy*. Oxford: Oxford University Press, 2001.

Verba, Sidney. 'Sequences and Development'. In *Crises and Sequences in Political Development*, edited by Leonard Binder. Princeton: Princeton University Press, 1971.

Vogel, David. *Fluctuating Fortunes: The Political Power of Business in America*. New York: Basic Books, 1989.

Volkov, Vadim. *Violent Entrepreneurs: The Use of Force in the Making of Russian Capitalism*. Ithaca: Cornell University Press, 2002.

Wallin, Thomas E. *West Siberia: The Key to Russia's Oil and Gas Future*. New York: Petroleum Intelligence Group, 1992.

Waterbury, John. 'The Heart of the Matter? Public Enterprise and the Adjustment Process'. In *The Politics of Economic Adjustment: International Constraints, Distributive Conflicts and the State*, edited by Stephan Haggard and Robert R. Kaufman. Princeton: Princeton University Press, 1992.

——. 'The Political Management of Economic Adjustment and Reform'. In *Fragile Coalitions: The Politics of Economic Adjustment*, edited by Joan M. Nelson. New Brunswick: Transaction Books, 1989.

Wedel, Janine. *Collision and Collusion: The Strange Case of Western Aid to Eastern Europe*. New York: St Martin's Press, 1998.

Weiss, Linda. *The Myth of the Powerless State*. New York: Cornell University Press, 1998.

White, Stephen, Richard Rose and Ian McAllister. *How Russia Votes*. New Jersey: Chatham House, 1997.

Whitefield, Stephen. *Industrial Power and the Soviet State*. Oxford: Clarendon Press, 1993.

Williamson, John, ed. *The Political Economy of Policy Reform*. Washington DC: Institute for International Economics, 1994.

——. 'In Search of a Manual for Technopols'. In *The Political Economy of Policy Reform*. Washington DC: Institute for International Economics, 1994.

Williamson, Peter J. *Corporatism in Perspective: An Introductory Guide to Corporatist Theory*. London: SAGE Publications, 1989.

Wilson, Graham K. *Interest Groups*. Oxford: Basil Blackwell, 2001.

——. *Business and Politics: A Comparative Introduction*. Basingstoke: Palgrave Macmillan, 1990.

World Bank. *Economies in Transition: An OED Evaluation of World Bank Assistance*. Washington DC: World Bank, 2004.

——. *From Transition to Development: A Country Economic Memorandum for the Russian Federation*. Moscow: World Bank Group, Russian Federation Country Office, 2004.

——. *Russian Economic Report*. Moscow: World Bank Group, Russian Federation Country Office, November 2004.

——. *Russian Economic Report*. Moscow: World Bank Group, Russian Federation Country Office, February 2004.

Yel'tsin, Boris. *Midnight Diaries*. Translated by Catherine A. Fitzpatrick. New York: Public Affairs, 2000.

——. *The Struggle for Russia*. Translated by Catherine A. Fitzpatrick. New York: Random House, 1994.

——. *Against the Grain: An Autobiography*. Translated by Michael Glenny. New York: Summit Books, 1990.

Journal Articles

Almond, Gabriel A. 'Corporatism, Pluralism and Professional Memory'. *World Politics* 35, no. 2 (January 1983): 245–260.

Al-Othman, Abdullatif A. 'The Reliable Supplier'. *Foreign Affairs* 81, no. 6 (2002): 173–175.

Appel, Hilary. 'The Ideological Determinants of Liberal Economic Reform'. *World Politics* 52, no. 4 (July 2000): 520–549.

Aslund, Anders. 'Russia's Economic Transformation under Putin'. *Eurasian Geography and Economics* 45, no. 6 (2004): 397–420.

——. 'Reform vs "Rent Seeking" in Russia's Economic Transformation'. *Transition* (26 January 1996): 12–16.

Balzer, Harley. 'Managed Pluralism: Vladimir Putin's Emerging Regime'. *Post-Soviet Affairs* 19, no. 3 (2003): 189–227.

Bartle, Ian. 'When Institutions No Longer Matter: Reform of Telecommunications and Electricity in Germany, France and Britain'. *Journal of Public Policy* 22, no. 1 (2002): 1–27.

Baumgartner, Frank R. and Bryan D. Jones. 'Agenda Dynamics and Policy Subsystems'. *The Journal of Politics* 53, no. 4 (1991): 1044–1074.

Becker, Gary S. 'A Theory of Competitions among Pressure Groups for Political Influence'. *The Quarterly Journal of Economics* 48, no. 3 (August 1983): 371–400.

Benoit, Kenneth and Jacqueline Hayden. 'Institutional Change and Persistence: The Evolution of Poland's Electoral System, 1989–2001'. *The Journal of Politics* 66, no. 2 (2004): 1468–2508.

—— and J. W. Schleimann. 'Institutional Choice in New Democracies: Bargaining over Hungary's 1989 Electoral Law'. *Journal of Theoretical Politics* 13, no. 2 (2001): 153–182.

Bhagwati, Jagdish N. 'Directly Unproductive, Profit-Seeking (DUP) Activities'. *Journal of Political Economy* 90, no. 5 (1982): 988–1002.

Birdsall, Nancy and Arvind Subramaniam. 'Saving Iraq from Its Oil'. *Foreign Affairs* 83, no. 4 (2004): 77–89.

Black, Bernard, Reiner Kraakman and Anna Tarassova. 'Russian Privatisation and Corporate Governance: What Went Wrong?' *Stanford Law Review* 52 (2000): 1731–1808.

Blum, Douglas W. 'Domestic Politics and Russia's Caspian Policy'. *Post-Soviet Affairs* 14, no. 2 (1998): 137–164.

Brada, Joseph C. 'Privatisation is Transition – Or Is It?' *Journal of Economic Perspectives* 10, no. 2 (1996): 67–86.

Bradshaw, Michael and Andrew R. Bond. 'Crisis Amid Plenty Revisited: Comments on the Problematic Potential of Russian Oil'. *Eurasian Geography and Economics* 45, no. 5 (2004): 352–358.

Breslauer, George W. 'Boris Yel'tsin as Patriarch'. *Post-Soviet Affairs* 15, no. 2 (1999): 186–200.

——. 'Soviet Economic Reforms since Stalin: Ideology, Politics and Learning'. *Soviet Economy* 6, no. 3 (1990): 252–280.

Brovkin, Vladimir. 'Wishful Thinking about Russia?'. *Transition* (June 1999): 22–25.

Brown, Archie. 'Vladimir Putin and the Reaffirmation of Central State Power'. *Post-Soviet Affairs* 17, no. 1 (2001): 45–55.

Brown, J. David and John S. Earle. 'Evaluating Enterprise Privatisation in Russia'. *Russian Economic Trends*, no. 3 (1999).

Bruno, Michael. 'Stabilisation and the Macroeconomics of Transition: How Different is Eastern Europe?'. *Economics of Transition* (1993): 5–19.

Bruszt, Laszlo and David Stark. 'Postcommunist Networking: Secret Agents, Mafiosi and Sociologists'. *East European Constitutional Review* (2000): 115–120.

Brzezinski, Zbigniew. 'The Primacy of History and Culture'. *Journal of Democracy* 12, no. 3 (2001): 21–26.

Chaisty, Paul and Petra Schleiter. 'Productive But Not Valued: The Russian State Duma, 1994–2001'. *Europe–Asia Studies* 55, no. 5 (2002): 701–724.

Clark, William Roberts. 'Agents and Structures: Two Views of Preferences, Two Views of Institutions'. *International Studies Quarterly* 42, no. 2 (1998): 245–270.

Dallin, Alexander. 'Causes of the Collapse of the USSR'. *Post-Soviet Affairs* 8, no. 4 (1992): 279–302.

Davis, Jeffrey, Rolando Ossowski, James Daniel and Steven Barnett. 'Oil Funds: Problems Posing as Solutions?' *Finance & Development* 38, no. 4 (2001). Available online from http://www.imf.org).

de Melo, Martha, Cevdet Denizer, Alan Gelb and Stoyan Tenev. 'Circumstance and Choice: The Role of Initial Conditions and Policies in Transition Economies'. *The World Bank Economic Review* 15, no. 1 (2001): 1–31.

Dienes, Leslie. 'The Energy System and Economic Imbalances in the USSR'. *Soviet Economy* 1, no. 4 (1985): 340–372.

——. 'Observations on the Problematic Potential of Russian Oil and the Complexities of Siberia'. *Eurasian Geography and Economics* 45, no. 5 (2004): 319–345.

Dowding, Keith. 'The Compatability of Behaviouralism, Rational Choice and 'New Institutionalism' *Journal of Theoretical Politics* 6, no. 1 (1994): 105–117.

Dunleavy, Patrick. 'Explaining the Privatisation Boom: Public Choice versus Radical Approaches'. *Public Administration* 64, no. 1 (1986): 13–34.

Easter, Gerald M. 'Politics of Revenue Extraction in Post-Communist States: Poland and Russia Compared'. *Politics & Society* 30, no. 4 (2002): 599–627.

Fairbanks, Charles H. 'The Feudalisation of the State'. *Journal of Democracy* 10, no. 2 (1999): 47–53.

Feigenbaum, Harvey B. and Jeffrey R. Henig. 'The Political Underpinnings of Privatisation: A Typology'. *World Politics* 46, no. 2 (1994): 185–208.

Fischer, Stanley and Alan Gelb. 'The Process of Socialist Economic Transformation'. *Journal of Economic Perspectives* 5, no. 4 (1991): 91–105.

Frye, Timothy. 'A Politics of Institutional Choice: Post-Communist Presidencies'. *Comparative Political Studies* 30, no. 5 (October 1997): 523–552.

——. 'Capture or Exchange? Business Lobbying in Russia'. *Europe–Asia Studies* 54, no. 7 (2002): 1017–1036.

—— and Andrei Shleifer. 'The Invisible Hand and the Grabbing Hand'. *The American Economic Review: Papers and Proceedings of the 100th Annual Meeting of the American Economic Association* 87, no. 2 (1997): 354–358.

Gaddy, Clifford and Barry Ickes. 'Russia's Virtual Economy'. *Foreign Affairs* 77, no. 5 (1998): 53–67.

——. 'Perspectives on the Potential of Russian Oil'. *Eurasian Geography and Economics* 45, no. 5 (2004): 346–351.

Gaydar, Yegor. 'Vosstanovitel'nyy rost i nekotorye osobennosti sovremennoy ekonomicheskoy situatsii v Rossii'. *Voprosy ekonomiki*, no. 6 (2003): 4–18.

Galbraith, John Kenneth. 'Countervailing Power'. *American Economic Review: Papers and Proceedings of the 66th Annual Meeting of the American Economic Association* 44 (1954): 1–6.

Goldman, Marshall I. 'Anders in Wonderland: Comments on Russia's Economic Transformation under Putin'. *Eurasian Geography and Economics* 45, no. 6 (2004): 429–434.

Goltz, Thomas. 'Letter from Eurasia: The Hidden Russian Hand'. *Foreign Policy*, no. 92 (1993): 92–116.

Grant, Jordan A. 'Pluralistic Corporatisms and Corporate Pluralism'. *Scandinavian Political Studies* 7, no. 3 (1984): 137–161.

——. 'Iron Triangles, Woolly Corporatism and Elastic Nets: Images of the Policy Process'. *Journal of Public Policy* 1, no. 1 (1981): 95–123.

Haas, Peter M. 'Introduction: Epistemic Communities and International Policy Coordination'. *International Organisation* 46, no. 2 (1992): 1–35.

Hahn, Gordon M. 'The Impact of Putin's Federative Reforms on Democratisation in Russia'. *Post-Soviet Affairs* 19, no. 2 (2003): 114–153.

Hall, Peter A. 'Policy Paradigms, Social Learning and the State: The Case of Economic Policymaking in Britain'. *Comparative Politics* 3, no. 25 (1993): 275–296.

—— and Rosemary C. R. Taylor. 'Political Science and the Three New Institutionalisms'. *Political Studies* XLIV, no. 5 (1996): 936–957.

Hanson, Philip. 'Putin and Russia's Economic Transformation'. *Eurasian Geography and Economics* 45, no. 6 (2004): 421–428.

Harsanyi, John C. 'Rational Choice Models of Political Behaviour vs. Functionalist and Conformist Theories'. *World Politics* 21, no. 4 (1969): 513–538.

Hellman, Joel S. 'Winners Take All: The Politics of Partial Reform in Postcommunist Transitions'. *World Politics* 50, no. 2 (1998): 203–234.
—— and Daniel Kaufmann. 'Confronting the Challenge of State Capture in Transition Economics'. *Finance & Development* 38, no. 3 (2001): 1.
Hill, Fiona and Florence Fee. 'Fueling the Future: The Prospects for Russian Oil and Gas'. *Democratisation* 10, no. 4 (2002): 462–487.
Holmes, Stephen. 'Superpresidentialism and its Problems'. *East European Constitutional Review* 3, no. 1 (1994): 123–126.
Holt, David, David A. Ralston and Robert H. Terpstra. 'Constraints on Capitalism in Russia: The Managerial Psyche, Social Infrastructure, and Ideology'. *California Management Review* 36, no. 3 (1994): 124–141.
Jacobsen, John Kurt. 'Much Ado about Ideas: The Cognitive Factor in Economic Policy'. *World Politics* 47, no. 2 (1995): 283–310.
Jaffe, Amy Meyers and Robert A. Manning. 'Russia, Energy and the West'. *Survival* 43, no. 2 (Summer 2001): 133–152.
Johnson, Juliet. 'Russia's Emerging Financial-Industrial Groups'. *Post-Soviet Affairs* 13, no. 4 (1997): 333–365.
Johnson, Simon and Heidi Kroll. 'Managerial Strategies for Spontaneous Privatisation'. *Soviet Economy* 7 (1991): 281–316.
Kaufmann, Daniel and Paul Siegelbaum. 'Privatisation and Corruption in Transition Economies'. *Journal of International Affairs* 50, no. 2 (1996): 422–458.
Khartukov, Eugene and Ellen Starostina. 'Ex-Soviet Oil Exports: Are the Russians Really Coming?' *Middle East Economic Survey* XLVII, no. 4 (26 January 2004). Available online from http://www.mees.com/postedarticles/oped/a47n04d01.htm.
Krasner, Stephen D. 'Approaches to the State: Alternative Conceptions and Historical Dynamics'. *Comparative Politics* 16, no. 2 (January 1984): 223–246.
Kryshtanovskaya, Olga and Stephen White. 'From Soviet Nomenklatura to Russian Elite'. *Europe–Asia Studies* 48, no. 5 (1996): 711–733.
——. 'Putin's Militocracy'. *Post-Soviet Affairs* 19, no. 4 (2003): 289–306.
Kwon, Goohoon. 'Post-crisis Revenue Developments in Russia: From an Oil Perspective'. *Public Finance and Management* 3, no. 4 (2003): 505–530.
Lane, David. 'The Transformation of Russia: The Role of the Political Elite'. *Europe–Asia Studies* 48, no. 4 (1996): 535–549.
Le Huerou, Anne. 'Elites in Omsk'. *Post-Soviet Affairs* 15, no. 4 (1999): 362–386.
Levin, Mark and Georgy Saratov. 'Corruption and Institutions in Russia'. *Journal of Political Economy* 16 (2000): 113–132.
Lieberman, Ira W. 'Privatisation: The Theme of the 1990s'. *The Columbia Journal of World Business* 28, no. 1 (1993): 9–15.
—— and Rogi Veimetra. 'The Rush for State Shares in the "Klondike" of Wild East Capitalism: Loans-for-Shares Transactions in Russia'. *The George Washington Journal of International Law and Economics* 29, no. 3 (1996): 737–768.
Lieberman, Robert C. 'Ideas, Institutions and Political Order: Explaining Political Change'. *American Political Science Review* 96, no. 4 (December 2002): 697–712.
Lijphart, Arend. 'Comparative Politics and Comparative Method'. *American Political Science Review* 65 (1971): 682–693.
——. 'The Comparable Cases Strategy in Comparative Research'. *Comparative Political Studies*, no. 8 (1975): 158–177.

Lowi, Theodore J. 'American Business, Public Policy, Case Studies and Political Theory'. *World Politics* 16, no. 4 (July 1964): 677–715.

Luong, Pauline Jones and Erika Weinthal. 'Prelude to the Resource Curse: Explaining Oil and Gas Development Strategies in the Soviet Successor States and Beyond'. *Comparative Political Studies* 34, no.4 (2001): 367–399.

Luthans, Fred, Dianne H. B. Welsch and Stuart A. Rosenkratz. 'What do Russian Managers Really Do? An Observational Study with Comparisons to US Managers'. *Journal of International Business Studies* 24, no. 4 (1993): 741–761.

Malleret, Thierry, Natalia Orlova and Vladimir Romanov. 'What Loaded and Triggered the Russian Crisis?' *Post-Soviet Affairs* 15, no. 2 (1999): 107–129.

March, James G. and Johan P. Olsen. 'The New Institutionalism: Organisational Factors in Political Life'. *American Political Science Review* 78, no. 3 (1984): 738–749.

Marsh, David and Martin Smith. 'Understanding Policy Networks: Towards a Dialectical Approach'. *Political Studies* 48 (2000): 4–21.

Martin, Ross. 'Pluralism and the New Corporatism'. *Political Studies*, no. 31 (1983): 86–102.

McCubbins, Matthew, Roger G. Noll and Barry R. Weingast. 'Structure and Process, Politics and Policy: Administrative Arrangements and the Political Control of Agencies'. *Virginia Law Review* 75 (1989): 431–483.

McFaul, Michael. 'Russia's "Privatised" State as an Impediment to Democratic Consolidation (Part I)'. *Security Dialogue* 29, no. 2 (1998): 191–199.

——. 'Russia's "Privatised" State as an Impediment to Democratic Consolidation (Part II)'. *Security Dialogue* 29, no. 2 (1998): 315–332.

——. 'State Power, Institutional Change and the Politics of Privatisation in Russia'. *World Politics* 47, no. 2 (January 1995): 210–243.

——. "Institutional Design, Uncertainty and Path Dependency during Transitions: Cases from Russia." *Constitutional Political Economy* 10, no. 1 (1999): 27–52.

Megginson, William L., Robert C. Nash and Matthias von Randenborgh. 'The Financial and Operating Performance of Newly Privatised Firms: An International Empirical Analysis'. *Journal of Finance* XLIX (1994): 403–452.

—— and Jeffrey M. Netter. 'From State to Market: A Survey of Empirical Studies on Privatisation'. *Journal of Economic Literature* 39, no. 2 (June 2001): 321–389.

Mendras, Marie. 'How Regional Elites Preserve their Power'. *Post-Soviet Affairs* 15, no. 4 (1999): 295–311.

Millar, James R. 'Papa Schaeg on Economic Reform in Russia'. *Problems of Post-Communism* 48, no. 3 (2001): 3–9.

Mishler, William and John P. Willerton. 'The Dynamics of Presidential Popularity in Post-Communist Russia: Cultural Imperative versus Neo-Institutional Choice?' *The Journal of Politics* 65, no. 1 (2003): 111–141.

Moore, Mick. 'Revenues, State Formation, and the Quality of Governance in Developing Countries'. *International Political Science Review* 25, no. 3 (2004): 297–319.

Morse, Edward L. and James Richard. 'The Battle for Energy Dominance'. *Foreign Affairs* 81, no. 2 (2002): 16–31.

Murphy, K., Andrei Shleifer and Robert W. Vishny. 'Why is Rent-Seeking so Costly to Growth?' *American Economic Review* 83, no. 3 (1993): 409–414.

Naim, Moises. 'Russia's Oily Future'. *Foreign Policy*, no. 140 (2004): 95–96.

Nash, Roland and Dirk Willer. 'Share Prices in Russia: The Reasons for Undervaluation'. *Russian Economic Trends* 4, no. 2 (1995): 111–126.

Niskanen, William A. 'Bureaucrats and Politicians'. *Journal of Law and Economics*, no. 18 (1975): 617–643.

North, Douglass C. 'A Framework for Analysing the State in Economic History'. *Explorations in Economic History*, no. 16 (1979): 249–259.

Ostrom, Elinor. 'Rational Choice Theory and Institutional Analysis: Towards Complementarity'. *American Political Science Review* 85, no. 1 (1991): 237–243.

Pappe, Yakov. 'Neftyanaya i gazovaya diplomatiya Rossii'. *Pro et Contra*, no. 3 (1997). Available online from http://www.carnegie.ru.

Peregudov, Sergei and Irina Semenenko. 'Lobbying Business Interests in Russia'. *Democratisation* 3, no. 2 (1996): 115–139.

Peregudov, Sergei. 'Krupnaya rossiyskaya korporatsiya v sisteme vlasti'. *Polis*, no. 3 (2001): 21–33.

Pipes, Richard. 'Flight from Freedom'. *Foreign Affairs* 83, no. 3 (2004): 9–15.

Puffer, Sheila M. 'Shedding the Legacy of the Red Executive: Leadership in Russian Enterprises'. *International Business Review* 4, no. 2 (1995): 157–176.

Reddaway, Peter. 'Is Putin's Power more Formal than Real?' *Post-Soviet Affairs* 18, no. 1 (January–March 2002): 31–40.

Remington, Thomas F. 'Putin's Third Way: Russia and the "Strong State" Ideal'. *East European Constitutional Review*, 9, no. 1/2 (2000). Available online from http://www.law.nyu.edu/eecr.

Reynolds, Douglas B. 'Soviet Economic Decline: Did an Oil Crisis Cause the Transition in the Soviet Union?' *The Journal of Energy and Development* 24, no. 1 (2000): 65–81.

Roberts, Cynthia and Thomas Sherlock. 'Bringing the State Back In: Explanations of the Derailed Transition to Market Democracy'. *Comparative Politics* 31, no. 4 (1999): 477–498.

Robinson, Neil. 'The Myth of Equilibrium: Winner Power, Fiscal Crisis and Russian Economic Reform'. *Communist and Post-Communist Studies* 34, no. 4 (2001): 423–446.

Rosefielde, Steven. 'Why Policymakers Don't Listen'. *Problems of Post-Communism* 48, no. 3 (2001): 4–5.

Rumer, Boris. 'Structural Imbalance in the Soviet Economy'. *Problems of Communism* 33 (1984): 24–32.

Rutland, Peter. 'Privatisation in Russia: One Step Forward: Two Steps Back?' *Europe–Asia Studies* 46, no. 7 (1994): 1109–1131.

——. 'Putin's Path to Power'. *Post-Soviet Affairs* 16, no. 4 (2000): 313–354.

——, Gail W. Lapidus, Barry Ickes, Carol Saivetz and George W. Breslauer. 'Russia in the Year 2003'. *Post-Soviet Affairs* 20, no. 1 (2004): 1–45.

Sagers, Matthew J. 'The Russian Natural Gas Industry in the Mid-1990s'. *Post-Soviet Geography and Economics* 36, no. 9 (1995): 521–564.

Sappington, David. 'Incentives in Principal–Agent Relationships'. *Journal of Economic Perspectives* 3, no. 2 (1991): 45–66.

Schamis, Hector E. 'Distributional Coalitions and the Politics of Economic Reform in Latin America'. *World Politics* 51 (1999): 236–268.

Schlager, Edella and William Blomquist. 'A Comparison of Three Emerging Theories of the Policy Process'. *Political Research Quarterly* 49 (1996): 651–672.

Schröder, Hans-Henning. 'El'tsin and the Oligarchs: The Role of Financial Groups in Russian Politics between 1993 and July 1998'. *Europe–Asia Studies* 51, no. 6 (1999): 957–988.

Schroeder, Gertrude E. 'The Slowdown in Soviet Industry, 1976–1982'. *Soviet Economy* 1, no. 1 (1985): 42–74.

Sestanovich, Stephen. 'Force, Money, and Pluralism'. *Journal of Democracy* 15, no. 3 (2004): 32–42.

Shepsle, Kenneth A. 'Studying Institutions: Some Lessons from the Rational Choice Approach'. *Journal of Theoretical Politics* 11, no. 2 (1989): 131–147.

Shevtsova, Lilia. 'The Limits of Bureaucratic Authoritarianism'. *Journal of Democracy* 15, no. 3 (2004): 67–77.

Shkaratan, Ovsei and Yuriy Figatner. 'Starye i novye khozyaeva Rossii'. *Mir Rossii* 1, no. 1 (1992): 67–90.

Shlapentokh, Vladimir. 'Wealth versus Political Power: The Russian Case'. *Communist and Post-Communist Studies* 37, no. 2 (2004): 135–160.

——. 'Early Feudalism: The Best Parallel for Contemporary Russia'. *Europe–Asia Studies* 48, no. 3 (1996): 392–412.

——. 'Privatisation Debates in Russia, 1989–1992'. *Comparative Economic Studies* 35, no. 2 (1993): 19–32.

Shleifer, Andrei and Robert W. Vishny. 'Politicians and Firms'. *The Quarterly Journal of Economics* 109, no. 4 (1994): 995–1025.

Shleifer, Andrei and Daniel Treisman. 'A Normal Country'. *Foreign Affairs* 83, no. 2 (2004): 20–38.

Simon, Herbert A. 'Organisations and Markets'. *Journal of Economic Perspectives* 3, no. 2 (1991): 25–44.

——. 'A Behavioural Model of Rational Choice'. *The Quarterly Journal of Economics*, LXIX (1955): 99–118.

Slay, Ben and Vladimir Capelik. 'The Struggle for Natural Monopoly Reform in Russia'. *Post-Soviet Geography and Economics* 38, no. 7 (1997): 396–429.

Stark, David. 'Path Dependence and Privatisation Strategies in East Central Europe'. *East European Politics and Society* 6, no. 1 (1992): 17–54.

Stewart, Philip D. 'Soviet Interest Groups and the Policy Process: The Repeal of Production Education'. *World Politics* 29, 1969: 29–51.

Stowe, Robert. 'Foreign Policy Preferences of the New Russian Business Elite'. *Problems of Post-Communism* 48, no. 3 (2001): 49–58.

Szegvari, Ivan. 'Who is to Blame for Russia's Economic Woes?' *Transition* (April 1999): 26–28.

Telegina, Yelena. 'Mirovoy energeticheskiy rynok i geopoliticheskie interesy Rossii'. *Mirovaya Ekonomika i Mezhdunarodnye Otnosheniya*, no. 5 (2003): 60–64.

Telhami, Shibley and Fiona Hill. 'America's Vital Stakes in Saudi Arabia'. *Foreign Affairs* 81, no. 6 (2002): 167–173.

Tompson, William. 'Putin's Challenge: The Politics of Structural Reform in Russia'. *Europe–Asia Studies* 54, no. 6 (2002): 933–957.

——. 'Was Gaidar Really Necessary?' *Problems of Post-Communism* 49, no. 4 (1992): 12–21.

Tikhomorov, Vladimir. 'Capital Flight from Post-Soviet Russia'. *Europe–Asia Studies* 49, no. 4 (1997): 591–615.

Treisman, Daniel. 'Russia Renewed?' *Foreign Affairs* (November/December 2002): 58–72.

Vogel, David. 'The Power of Business in America: A Re-Appraisal'. *British Journal of Political Science* 13, no. 1 (1983): 19–43.

Weingast, Barry R. 'A Rational Choice Perspective on the Role of Ideas: Shared Belief Systems and State Sovereignty in International Co-operation'. *Politics & Society* 23, no. 4 (1995): 449–464.

Wilsford, David. 'Path Dependency or Why History Makes it Difficult but Not Impossible to Reform Health Care Systems in a Big Way'. *Journal of Public Policy* 14, no. 3 (1994): 251–283.

Wollman, Hellmut. 'Change and Continuity of Political and Administrative Elites in Post-Communist Russia'. *Governance*, no. 6 (1993): 326–340.

Woodruff, David M. 'Rules for Followers: Institutional Theory and the New Politics of Economic Backwardness in Russia'. *Politics & Society* 28, no. 4 (2000): 437–482.

——. 'It's Value that's Virtual: Bartles, Rubles, and the Place of Gazprom in the Russian Economy'. *Post-Soviet Affairs*, 15, no. 2 (1999): 130–148.

——. 'Property Rights in Context: Privatisation's Legacy for Corporate Legality in Poland and Russia'. *Studies in Comparative International Development* 38, no. 4 (2004): 82–106.

Yee, Albert. 'The Causal Effects of Ideas on Policies'. *International Organisation* 50, no. 4 (1996): 69–108.

Yorke, Andrew. 'Business and Politics in Krasnoyarsk Krai'. *Europe–Asia Studies* 55, no. 2 (2003): 241–262.

Zudin, Aleksey. 'Neo-koporativizm v rossiya? Gosudarstvo i biznes pri Vladimire Putine'. *Pro et Contra* 6, no. 4 (Fall 2001): 171–198.

Working/Conference Papers

Algieri, Bernardina. 'The Effects of the Dutch Disease in Russia'. In *Zentrum fur Entwicklungsforschung (ZEF) Discussion Papers on Development Policy*, Bonn, January 2004.

Artemiev, Igor and Michael Haney. 'The Privatisation of the Russian Coal Industry: Policies and Processes in the Transformation of a Major Industry'. In *World Bank Transition Economies Working Paper*, Washington DC: World Bank, April 2002.

Aslund, Anders. 'Why Has Russia's Economic Transformation Been So Arduous?' Paper presented as the Annual World Bank Conference of Development Economics, Washington DC, 28–30 April 1999.

Baev, Pavel. 'Putin Reconstitutes Russia's Great Power Status'. In *PONARS Policy Memo*, Washington DC: Centre for Strategic and International Studies, November 2003.

Belin, Laura. 'The Fall and Rise of State Power over the Russian Media, 1995–2001'. D.Phil. thesis, Oxford University, May 2002.

Bobylev, Youri and Jacek Cukrowski. *Russia: Bank Assistance for the Energy Sector*. In *Operations Evaluation Department Working Paper*, Washington DC: World Bank, 2002.

Boone, Peter and Denis Rodionov. 'Rent Seeking in Russia and the CIS'. Paper presented at the Tenth Anniversary Conference of the European Bank for Reconstruction and Development, London, December 2001.

Broadman, Harry G. *Case-by-Case Privatisation in Russia: Principles and Institutions*. Washington DC: World Bank, 1998.

Desai, Raj M and Itzhak Goldberg. 'The Vicious Circles of Control: Regional Governments and Insiders in Privatised Russian Enterprises'. In *World Bank Working Paper*, Washington DC: World Bank, 2000.

Djankov, Simeon and Peter Murrell. *The Determinants of Enterprise Restructuring in Transition: An Assessment of the Evidence*. Washington DC: World Bank, September 2000.

Faccio, Mara. 'Politically Connected Firms'. *Owen Graduate School of Management Working Paper*, September 2003.

Fedorov, Boris. 'Russia after the Presidential Election'. Talk held at the Carnegie Endowment for International Peace, Washington DC, 30 March 2004.

Frye, Timothy. 'Slapping the Grabbing Hand: Credible Commitment, State Capacity and Property Rights in Russia'. Unpublished manuscript.

Isham, Jonathan, Michael Woolcock, Lant Pritchett and Gwen Busby. 'The Varieties of Resource Experience: How Natural Resource Export Structures Affect the Political Economy of Economic Growth'. In *Rohatyn Centre for International Affairs*, Working Paper no.12, 2003.

Kabalina, Veronika, Vadim Borisov, Simon Clarke and Peter Fairbrother. 'Privatisation and the Struggle for Control: The Case of Coal Mining'. In *Russian Research Programme*, Warwick University, June 1994.

Kim, S. Ran and A. Horn. 'Regulating Policies concerning Natural Monopolies in Developing and Transition Economies'. In *DESA Discussion Paper no. 8*, March 1999.

Lane, Philip and Aaron Tornell. 'Power, Concentration and Growth'. In *Harvard Institute of Economic Research Discussion Paper*, Harvard, May 1995.

Lindert, Peter H. 'Voice and Growth: Was Churchill Right?' Working Paper 109, Davis: University of California, October 2002.

Mau, Vladimir. 'Russian Economic Reforms as Perceived by Western Critics'. *Bank of Finland Institute for Economies in Transition (BOFIT) Discussion Papers*, Helsinki, December 1999.

Moser, Nat. 'The Privatisation of the Russian Oil Industry 1992–1995: Façade or Reality?' M.Phil. thesis, Oxford University, 1996.

Nellis, John. *The World Bank, Privatisation and Enterprise Reform in Transition Economies: A Retrospective Analysis*. World Bank Working Paper, Washington DC: World Bank, January 2002.

North, Douglass C. 'The Contribution of the New Institutional Economics to an Understanding of the Transition Problem'. Paper presented at the WIDER Annual Lectures 1, Helsinki March 1997.

Odling-Smee, John. 'The IMF and Russia in the 1990s'. In *IMF Working Paper*, Washington DC, August 2004.

Peregudov, Sergei. 'Business Interest Groups in the Political Systems of USSR and Russia'. Warwick: European Public Policy Institute, 1994.

Peters, B. Guy and Brian W. Hogwood. 'Policy Succession: The Dynamics of Policy Change'. In *Studies in Public Policy*. Glasgow: University of Strathclyde, 1980.

Piontkovsky, Andrei. *Considering Russia's Upcoming Elections: Putin's Challenges and Opportunities*, Talk given at the Centre for Strategic & International Studies, Washington DC, 11 February 2003.

Radygin, Alexander. 'Ownership and Control of the Russian Industry'. Paper presented at the OECD Conference on Corporate Governance in Russia, Moscow, 31 May–2 June 1999.

——. 'Corporate Governance through the Banks: The Experience in Russia'. Paper presented at the Eleventh Plenary Session of the OECD Advisory Group on Privatisation, Rome, 18–19 September 1997.

Rautava, Jouko. 'The Role of Oil Prices and the Real Exchange Rate in Russia's Economy'. In *Bank of Finland Institute for Economies in Transition (BOFIT) Discussion Papers*. Helsinki, 2002.

Reddaway, Peter. 'The Evolution of the Distribution of Political Power in Russia, 1990–2000'. In *Project on Systematic Change and International Security in Russia and the New States of Eurasia*. Washington DC: Johns Hopkins School of Advanced International Studies, 2000.

Robinson, Neil. 'Communist and Post-Communist States and Path Dependent Political Economy'. In *Limerick Papers in Politics and Public Administration*. Limerick: University of Limerick, 2003.

Rosefielde, Steven. 'An Abnormal Country'. In *Bank of Finland Institute for Economies in Transition (BOFIT) Discussion Papers*. Helsinki, 2004.

Stiglitz, Joseph E. 'Whither Reform? Ten Years of Transition'. Paper presented at the Annual World Bank Conference on Development Economics, Washington DC 28–30 April 1999.

Sprenger, Carsten. 'Ownership and Corporate Governance in Russian Industry: A Survey'. In *European Bank for Reconstruction and Development Working Paper*, January 2002.

Taylor, Brian D. Putin's State Building Project: Issues for the Second Term. In *PONARS Policy Memo*. Washington DC: Centre for Strategic and International Studies, November 2003.

Volkov, Vadim. 'The Yukos Affair: Terminating the Implicit Contract'. In *PONARS Policy Memo*. Washington DC: Centre for Strategic and International Studies, November 2003.

——. 'The Selective Use of State Capacity in Russia's Economy: Property Disputes and Enterprise Takeovers after 2000'. In *PONARS Policy Memo*. Washington DC: Centre for Strategic and International Studies, October 2002.

Woodruff, David M. 'Khodorkovsky's Gamble: From Business to Politics in the YUKOS Conflict'. In *PONARS Policy Memo*. Washington DC: Centre for Strategic and International Studies, November 2003.

——. 'The Extremely Hostile Takeover: Russian Lessons on Law, Property, and Politics'. Paper presented at the Annual Conference of the American Association for the Advancement of Slavic Studies, Toronto, Canada November 2003.

Newspapers/Weekly Periodicals

In Russian:

Ekonomika i zhizn'
Ekspert
Gazeta.ru
Izvestiya
Kommersant'-Daily

Kommersant'-Vlast'
Moskovskiye novosti
Nezavisimaya gazeta
Neftegazovaya vertikal'
Neft' i Kapital
Novaya gazeta
Profil'
Segodnya
Vedomosti
Vremya novostey
Yezhenedel'nyy zhurnal

In English:
Alexander's Oil and Gas Connections
Business Week
CDI Weekly
Current Digest of the Post-Soviet Press
The Economist
Financial Times (London)
Financial Times Energy Newsletters: East European Energy Report
International Herald Tribune
Johnson's Russia's List
Moscow News
The Moscow Times
The New York Times
Oil & Gas Journal
Petroleum Economist
Platt's Oilgram News
Prism
Radio Free Europe/Radio Liberty
Wall Street Journal
The Washington Post

Index

Abramovich, Roman, 47, 67, 91–3, 95, 116, 122, 126
  see also Sibneft'
actors, individual, 9–10, 24–30, 47–8, 55, 71–2, 96–7, 128, 130, 158n45
Agrarian Party, 43
agro-industrial sector, 22, 30
Alekperov, Vagit, 4, 6, 17–19, 50–1, 53, 55, 71, 120, 123, 125–7
  see also LUKoil
Al'fa bank/group, 26, 39–41, 56, 59–60, 63, 70, 82, 85–7, 91, 93, 124
  see also Fridman, Mikhail
Andropov, Yuriy, 19
Angarsk oil refinery, 39, 69
Angarsk pipeline project, 69–70
Anti-Monopoly Committee, 13, 23, 62, 64
Asian financial crisis, 83–4, 109
asset-stripping, 53–5, 72
auctions
  collateral, 37, 43, 61–5
  manipulation of, 58–9, 61–3, 76, 82, 91, 95, 107
  organization, 58–63, 91, 95, 109, 159n48
  specialized, 81
  timing of, 83, 94
  trust, 31, 58–60
  see also loans-for-shares scheme
Audit Chamber, 87–8, 124
Aushev, Ruslan, 89–90
Aven, Pyotr, 26–7

bandit, 'roving' and 'stationary', 55, 73
  see also Olson, Mancur
banking sector, 5, 8, 33–6, 46–7, 56–7, 59–60, 62–3, 122, 158n41
  see also individual banks
Bank Yugra, 77

Baranovskiy, Anatoliy, 92–3
Bashneft', 1, 39, 181n21
Belarus, 77, 79–81, 89, 95, 97, 123, 153n144
Berezovskiy, Boris, 29, 40, 57, 63, 102, 104, 106–7, 112, 124, 152n141, 153n143, 175n24
  see also Sibneft'
Bespalov, Yuriy, 103–7
BIN group, 87–9, 130
  see also Gutseriev, Mikhail
Bogdanchikov, Sergey, 68, 93, 110–15, 118, 121, 127
  see also Rosneft'
Bogdanov, Vladimir, 55, 120, 125–7
  see also Surgutneftegaz
bribery, 6, 46, 58, 90, 159n45, 47
British Petroleum (BP), 6, 20, 32, 70, 96, 110, 129, 180n11
budget-maximization, 24–5
bureaucratic agencies, 8, 12, 15, 24–6, 53, 63, 71, 74, 107, 118–20, 128, 140n77
  see also individual agencies

capture, privatization as, 3–4, 7, 14, 119, 121–5
cash privatization, 30, 56, 58, 158n36
Central Bank, 26, 58, 60, 63, 120
Chechnya, 84, 89–91, 111–12, 130, 170n64, 171n65, 177n51
Chernomyrdin, Viktor, 4, 13, 18, 22, 47, 65, 78–80, 101–8, 123, 176n36
Chubays, Anatoliy, 8, 22–3, 27–8, 42–3, 45, 64–5, 75, 81, 105–6, 109, 112, 123, 154n163, 175n31
  see also GKI; 'young reformers'
Churilov, Lev, 49–50, 71, 100
CIS/former republics, 79–81, 97, 123
coalition management strategies, 36–7, 41, 95–7, 114, 121–2, 152n128